A Mathematical Prelude to the Philosophy of Mathematics

Stephen Pollard

A Mathematical Prelude
to the Philosophy
of Mathematics

 Springer

Stephen Pollard
Department of Philosophy and Religion
Truman State University
Kirksville, MO
USA

ISBN 978-3-319-34833-9   ISBN978-3-319-05816-0   (eBook)
DOI 10.1007/978-3-319-05816-0
Springer Cham Heidelberg New York Dordrecht London

Printed on acid-free paper

Springer is part of Springer Science+Business Media (www.springer.com)

In memory of Dean Randall Pollard

φιλοσοφοῦμεν ἄνευ μαλακίας

Preface

I have a principled argument for why this book should exist. I have no such argument for why it contains just what it contains. The principles are these:

- You cannot understand philosophy of mathematics without understanding mathematics.
- You cannot understand mathematics without doing mathematics.

The main point of this book, with its 298 exercises, is to give students opportunities to recreate some mathematics that will illuminate important readings in philosophy of mathematics. As for the particular mathematical materials I have chosen: they are the unforced fruits of lengthy experience. I have taught undergraduates for three decades. In the 14 times I taught philosophy of mathematics, I discovered again and again that some important text was opaque to my students largely because they lacked some particular bit of mathematical background. Most of the missing bits came from a handful of subject areas: Primitive Recursive Arithmetic, Peano Arithmetic, Gödel's theorems (completeness, compactness, and incompleteness), interpretability, the hierarchy of sets (especially the landscape of $V(\omega + \omega)$), Frege Arithmetic, and intuitionist sentential logic. This book offers exercises in these areas supported by explanatory materials and just a dash of philosophy. I have made no attempt to impose a grand unifying narrative. There is no central thesis. Some students of philosophy of mathematics have found it helpful to work their way into these subject areas. Other readers may enjoy the same experience.

I cannot think of any undergraduate course in which this book would be an appropriate stand-alone text. I offer it as a supplement to primary texts chosen by instructors or automaths. For the benefit of the latter, the book offers some guidance about what those primary texts might be. Professors will, of course, make their own choices, both about primary texts and about assignments within this book. I would be amazed if any instructor assigned every section and every exercise. I do not see how there could be time for it. I would expect instructors to let their own interests guide them as they pick and choose. Not every choice will make sense. Chapter 2, for example, would be inscrutable without Chap. 1. Parts of Chap. 6 presuppose parts of Chap. 2. Chapters 4 and 5 build on Chap. 3. On the other hand, a leap-frog journey consisting of Chaps. 1, 3, and 7 would make sense—as would many other paths over and around sections instructors might choose to skip. That is one respect in which the book's somewhat motley character is a virtue.

The readers most likely to benefit from this book are exactly those most likely to benefit from a philosophy of mathematics course. Those readers will have some background in formal logic, they will find mathematics engaging and non-threatening, they will understand basic properties of the natural and real numbers, and they will see the point of asking "why" and "how" questions that emerge from mathematical experience, but cannot be answered by just producing more mathematics. I would describe this last trait as "philosophical inclination"—and it is the inclination, rather than any particular philosophical training, that is likely to be most important here. (Once, in a philosophy of social science course, a student of mine interrupted a discussion of deductive-nomological explanation by asking, "Why would anyone want to understand what explanation is?" There, I suspect, was a student who had never felt a philosophical impulse in his life. I would not recommend this book to him.)

I shared the joys and frustrations of a work-in-progress with my students. Some responded by dropping my course. Others stuck it out and helped make the book better. I like to think that, in return, they learned what it is like to *do* mathematics and philosophy. In any case, they have my thanks. Thanks, too, to Florence Emily Pollard for her careful reading of the penultimate draft.

Kirksville, MO, USA, January 2014   Stephen Pollard

Contents

1 Recursion, Induction 1
    1.1   Numerals, Types, Tokens 1
    1.2   Immediate Succession. 4
    1.3   How Many Types?. 7
    1.4   Recursive Definition. 9
    1.5   Proof by Induction 11
    1.6   A Characteristic Function 14
    1.7   Some Logic. 16
    1.8 Π_1^0 Sentences. 20
    1.9   Some Philosophy 24
    1.10   Solutions of Odd-Numbered Exercises 29
    References 33

2 Peano Arithmetic, Incompleteness 35
    2.1   The language of PA. 35
    2.2   The Axioms of PA. 37
    2.3   Incompleteness 1: Compactness. 39
    2.4   Incompleteness 2: Representability. 43
    2.5   Why Fret About Consistency? 50
    2.6   Solutions of Odd-Numbered Exercises 51
    References 52

3 Hereditarily Finite Lists 55
    3.1   What Sets Are Not. 55
    3.2   List-Types. 57
    3.3   Precursors. 60
    3.4   Logically Possible Lists 62
    3.5   Numbers Can Be Lists 64
    3.6   Lists Can Be Numbers 67
    3.7   Super-Numbers 70
    3.8   Hereditary Finiteness 72
    3.9   Finite Ranks 74

3.10 And Why Are We Doing This?   78
3.11 Solutions of Odd-Numbered Exercises   79
References   82

4   Zermelian Lists   85
    4.1 Infinite Lists   85
    4.2 More Ranks.   88
    4.3 Real Numbers   91
    4.4 From Lists to Sets   97
    4.5 Solutions of Odd-Numbered Exercises   97
    References   100

5   The Hierarchy of Sets   101
    5.1 New Axioms.   101
    5.2 Two Models   106
    5.3 How Many Ranks?.   107
    5.4 Equiconsistency.   109
    5.5 Even More Ranks   111
    5.6 $V(\omega + \omega)$ and Beyond   115
    5.7 Solutions of Odd-Numbered Exercises   118
    References   121

6   Frege Arithmetic   123
    6.1 The Language of FA   123
    6.2 The Axioms of FA.   126
    6.3 Some Number Theory   127
    6.4 FA Interprets PA.   135
    6.5 Extensions: An Epic Failure   141
    6.6 The Perils of Abstraction   143
    6.7 Monadic Frege Arithmetic   144
    6.8 Solutions of Odd-Numbered Exercises   153
    References   159

7   Intuitionist Logic   161
    7.1 Inference.   161
    7.2 Conjunctions   163
    7.3 Conditionals   166
    7.4 Negations   169
    7.5 Absurd Absurdities.   172
    7.6 Disjunctions   175
    7.7 Assimilators   178
    7.8 The Glivenko/Gödel Theorems   181
    7.9 First-Order Logic With a Decidable Relation   184

7.10   Curry's Paradox   187
7.11   Solutions of Odd-Numbered Exercises   188
References   196

Index   199

Chapter 1
Recursion, Induction

1.1 Numerals, Types, Tokens

The mathematician DAVID HILBERT (1862–1943) had the odd but, as it turned out, fertile idea that we can generate all sorts of interesting mathematics if we reflect on the *language* of mathematics—if we talk about mathematical talk. This is the magic of mathematics: start with a dry, unpromising little seed; nourish it with logic, imagination, and obsessive energy; if the mathematics gods smile, you find yourself wandering in a whole forest of beautiful ideas. Hilbert's little seed was a mathematical theory about the languages in which we express mathematical theories.

When we talk about what Hilbert did, we will be talking about talk about mathematical talk: we will be discussing a theory about the way mathematicians express themselves. We will do some of that in this chapter, *talking* about what Hilbert did. But, mainly, we will *do*, or re-do, what Hilbert did: we will experience a bit of Hilbert's project from the inside. This may require some patience on your part. We will be investigating an especially primitive bit of mathematical language in an especially meticulous way, working hard to prove elementary claims that you might have been perfectly happy to *assume*. The pay-off will be an insider's view of one of the most fundamental, most secure, and most deeply understood areas of mathematical activity: an area known as PRIMITIVE RECURSIVE ARITHMETIC or PRA. We begin our development of PRA by considering an archaic way of naming positive integers.

One way to answer a "How many?" question is to provide an *example* of how many. When the nice lady asks the little boy how old he is, the child holds up two fingers to show he is *that* many years old. When asked how old I am, I could, at the risk of appearing quite mad, display a piece of paper marked up like this:

IIIIIIIIIIIIIIIIIIIIIIIIIIIIIIIIIIIIIIIIIIIIIIIIIIIIIIIIIIIIIIIIIIII.

I could say, "Look: one mark for each year!" On each birthday, I could add a new tally mark. A year from now, my piece of paper would be marked up like this:

S. Pollard, *A Mathematical Prelude to the Philosophy of Mathematics*,
DOI: 10.1007/978-3-319-05816-0_1,
© Springer International Publishing Switzerland 2014

| | | | | | | | | | | | | | | | | | | | | | | | | | | | | | | | | | | | | | | | | | | | | | | | | | | | | | | | | | | | | | | | | | | | | | |.

I would then be employing a very simple system of NUMERATION, a very simple system for representing NUMBERS with NUMERALS.

Numerals are *names* of numbers. Numbers are the things *named* by numerals. The Roman numeral 'X' and the binary numeral '1010' are different numerals, but they name the same number. In my simple system, the numeral '|' names the number one, the numeral '||' names the number two, and so on. Each of my numerals *exemplifies* the number it names: a numeral consisting of n tally marks names the number n.

We now consider a distinction that is old hat to philosophers, but may not be so familiar to mathematicians. Suppose I hold up a piece of paper with pencil markings like this

$$| | | |$$

and you hold up a piece of paper with pencil markings like this

$$| | | |.$$

Then we are each displaying the *same* numeral: the one numeral in my simple system that names the number four; the one numeral consisting of four tally marks. There are two pieces of paper each bearing its own graphite inscriptions; but each graphite inscription is a particular instance of one and the same numeral. Now suppose you burn your piece of paper. Have you destroyed the numeral '||||'? I would be inclined to say that you have destroyed *a* numeral without destroying *the* numeral. Philosophers have some terminology that is useful here. They distinguish between the numeral-TYPE that survives the incineration of your paper and the numeral-TOKEN that goes up in flames. The type is one thing that has many instances or tokens. You can write one numeral a hundred times on a chalkboard. A hundred instances of the numeral will then appear on the board. Each of those instances is a token of the one type. You can destroy the hundred tokens by erasing the board. It is not so clear that you can do *anything* to destroy the one type.

For the rest of this chapter, when I refer to "numeral-tokens" and "numeral-types" I will mean tokens and types of numerals in our simple system of numeration. We now consider some properties of these tokens and types. Note, first, that each numeral-token is an instance of some numeral-type. To be a token is to be an instance of a type. To be a token is to "instantiate" a type. The very meaning of the word 'token' guarantees that there are no typeless tokens. The meaning does not guarantee that there *are* tokens or types. That would be an astounding thing for a meaning to do. The meaning only guarantees that if there were a token, there would be a type it instantiates. To take a more everyday example, the meanings of the terms 'wife' and 'spouse' do not guarantee that there are wives or spouses. They do guarantee that if there were a wife, she would have a spouse.

Each numeral-token is an instance of *at least* one numeral-type. Indeed, each numeral-token is an instance of *exactly* one numeral-type. The type of a numeral-token is determined by how many tally marks it includes. The graphite inscriptions on our two pieces of paper are tokens of the type: "numeral consisting of four tally marks." Neither inscription is a numeral consisting of three tally marks (though each has parts consisting of three tally marks). Neither inscription is a numeral consisting of five tally marks. In fact, each is an instance only of the type indicated. Since each numeral-token has a unique numeral-type, we can introduce some useful terminology.

Definition 1.1 If x is a numeral-token, we let $\tau(x)$ be x's numeral-type.

We can use our new symbol 'τ' (the Greek letter tau) to state an important proposition.

Proposition 1.1 *If x and y are numeral-tokens, then $\tau(x) = \tau(y)$ if and only if there are exactly as many tally marks in x as in y.*

Remember this proposition. It will very soon prove useful. It is known, in the technical language of philosophy, as an IDENTITY CRITERION for numeral-types. It supplies necessary and sufficient conditions for numeral-types (such as $\tau(x)$ and $\tau(y)$) to be identical to one another ($\tau(x) = \tau(y)$). At least, it does so for numeral-types that have tokens. We know that x is a token of $\tau(x)$. A numeral-token is, necessarily, a token of its own numeral-type. We are leaving open the possibility, however, that there are types with no tokens. There are no typeless tokens, but we are open to the idea that there are tokenless types. Our identity criterion would not apply to such types.

If we tried to formulate an identity criterion for numeral-*tokens*, we would face even thornier problems. Imagine two slips of paper, each bearing one tally mark. Bring the slips together to form a numeral-token y consisting of two tally marks. Do we have a clear idea of what changes to the slips would yield something other than y? What if we moved the slips one micron further apart? Would we still have y? If so, how many microns would we have to move the slips to get something other than y? What if we turned one of the slips upside down or applied a microscopic amount of additional graphite to it or moved the tally marks a tiny bit out of alignment? Would we still have y? We would have to think seriously about such questions before we could formulate an explicit identity criterion for numeral-tokens: a daunting task that I, frankly, am going to dodge. That does not leave us helpless. In this, as in so many areas, our capacity to make reasonable judgments exceeds our capacity to write down rules for making those judgments. We can make sensible decisions about the identity or distinctness of numeral tokens without formulating necessary and sufficient conditions for their identity or distinctness. It is worth remarking, though, that the mathematics we will develop in this chapter is emerging from notions infected with much more vagueness than we would tolerate in the mathematics itself.

1.2 Immediate Succession

We can construct longer and longer numeral-tokens by adding one tally mark at a time:

$$| \quad || \quad ||| \quad |||| \quad ||||| \quad |||||| \quad \cdots$$

If we add one tally mark to a numeral-token x, the only result of this construction that interests us, what we will call THE RESULT, is the new numeral-token consisting of all the tally marks of x together with the tally mark we added. (If any other tally marks have appeared or any of x's original tally marks have disappeared, then we have not successfully performed the intended operation.) We say that this result is what the construction "add one tally mark" YIELDS. If we add one tally mark at the end of a numeral-token x consisting of six tally marks, then one of the new objects we create is the inscription consisting of the last four tally marks in x together with the tally mark we added.

$$|| \underbrace{|||||}.$$

Even if we regard this inscription as a numeral-token (rather than just a part of a numeral-token) and concede that it is one result of our construction, it is not what we call *the result*. Our construction does not *yield* this numeral-token in our special sense of "yield."

In the next definition, we perform the everyday mathematical trick of compressing a lot of words into a little symbol: in this case, 'υ' (the Greek letter upsilon).

Definition 1.2 $\upsilon(xy)$ if and only if y is the result of adding one tally mark to numeral-token x.

Suppose y is the result of adding one tally mark to x while z is the result of adding one tally mark to w. That is, $\upsilon(xy)$ and $\upsilon(wz)$. Suppose, further, that $\tau(x) = \tau(w)$. Then, according to our identity criterion for numeral-types, x and w feature the same number of tally marks. If we add one tally mark to each, we will still have the same number of tally marks. So, applying our identity criterion again, $\tau(y) = \tau(z)$. We have confirmed the following entailment.

$$\tau(x) = \tau(w) \Longrightarrow \tau(y) = \tau(z).$$

Suppose, conversely, that $\tau(y) = \tau(z)$. This means that when we add a tally mark to x and a tally mark to w we get tokens with equal numbers of tally marks. So x and w must have equal numbers of tally marks and, hence, $\tau(x) = \tau(w)$. This confirms the converse of our earlier entailment.

$$\tau(y) = \tau(z) \Longrightarrow \tau(x) = \tau(w).$$

All of this establishes the following proposition.

Proposition 1.2 *If $\upsilon(xy)$ and $\upsilon(wz)$, then*

$$\tau(x) = \tau(w) \Longleftrightarrow \tau(y) = \tau(z).$$

Suppose x consists of two tally marks while y consists of three. There is no guarantee that $\upsilon(xy)$. Numeral-tokens can differ by one tally mark without either being a lengthening of the other. x might have been written on a chalkboard in Tuvalu fifty years ago and immediately erased, while y first appeared yesterday on a piece of paper in Missouri. Still, somewhere in the vast expanse of space-time, we would expect to find numeral-tokens w and z with the following properties.

$$\tau(w) = \tau(x), \quad \tau(z) = \tau(y), \quad \upsilon(wz).$$

We would expect to find a token consisting of three marks that *is* a lengthening of a token consisting of two marks. For example, z might be y itself and w might consist of the first two tally marks in y. These reflections inspire the following definition.

Definition 1.3 $\sigma(xy)$ if and only if a numeral-token of the same type as y is the result of adding one tally mark to a numeral-token of the same type as x.

The idea is that two numeral-tokens will stand in this relation σ ("sigma") if and only if the second token features exactly one more tally mark than the first (whether or not the second is the result of adding a tally mark to the first). We will now confirm some important facts about σ. Suppose $\sigma(xy)$. Then we can pick tokens x', y' with the properties we discussed above:

$$\tau(x') = \tau(x), \quad \tau(y') = \tau(y), \quad \upsilon(x'y').$$

Now suppose $\sigma(xz)$. Pick x'', z' such that

$$\tau(x'') = \tau(x), \quad \tau(z') = \tau(z), \quad \upsilon(x''z').$$

Then $\tau(x') = \tau(x'')$ and, hence, by Proposition 1.2, $\tau(y') = \tau(z')$. So $\tau(y) = \tau(z)$ and, on the assumption that $\sigma(xy)$, we have confirmed the following entailment.

$$\sigma(xz) \Longrightarrow \tau(y) = \tau(z).$$

Suppose, conversely, that $\tau(y) = \tau(z)$, still assuming that $\sigma(xy)$. Then $\tau(y') = \tau(z)$ and, hence, a numeral-token of the same type as z (namely y') is the result of adding one tally mark to a numeral-token of the same type as x (namely x'). So $\sigma(xz)$ and we have confirmed the converse of the earlier entailment.

$$\tau(y) = \tau(z) \Longrightarrow \sigma(xz).$$

Our reward is a new proposition.

Proposition 1.3 *If* $\sigma(xy)$, *then*

$$\sigma(xz) \Longleftrightarrow \tau(y) = \tau(z).$$

You can get warmed up by proving a closely related proposition.

Exercise 1.1 *Prove: if* $\sigma(xy)$, *then*

$$\sigma(zy) \Longleftrightarrow \tau(x) = \tau(z).$$

σ is a relation between numeral-*tokens*. There is a corresponding relation between numeral-*types*: the relation of IMMEDIATE SUCCESSION.

> 'II' immediately succeeds 'I'
> 'III' immediately succeeds 'II'
> 'IIII' immediately succeeds 'III'
> 'IIIII' immediately succeeds 'IIII'

And so on. This relation satisfies the following proposition.

Proposition 1.4 $\tau(y)$ *immediately succeeds* $\tau(x)$ *if and only if* $\sigma(xy)$.

Note that we are using a relation between tokens (the relation σ) to characterize a relation between types (the relation of immediate succession). We will now note some important facts about immediate succession. Suppose $\tau(y)$ and $\tau(z)$ both immediately succeed $\tau(x)$. Then, by Proposition 1.4, $\sigma(xy)$ and $\sigma(xz)$ and, hence, by Proposition 1.3, $\tau(y) = \tau(z)$. Say that a type is *instantiated* if there are tokens of that type. We have shown that an instantiated numeral-type will have at most one instantiated immediate successor. We are going to suppose that numeral-types behave like this whether or not they are instantiated.

Proposition 1.5 *Each numeral-type has at most one immediate successor.*

Now you need to do a little work to prepare the way for our next proposition.

Exercise 1.2 *Show that an instantiated numeral-type will immediately succeed at most one instantiated numeral-type.*

We are going to suppose that numeral-types behave as in Exercise 1.2 whether or not they are instantiated.

Proposition 1.6 *Each numeral-type immediately succeeds at most one numeral-type* (*so numeral-types that share an immediate successor are the same*).

In our system of numeration, at least one tally mark appears in each numeral-token. So you cannot add a tally mark to one of our numeral-tokens and end up with a single tally mark all by itself. Suppose $\tau(y) =$ ' I '. And suppose ' I ' immediately succeeds $\tau(x)$. Then, by Proposition 1.4, $\sigma(xy)$ and, hence, by Definition 1.3, a numeral-token of the same type as y is the result of adding one tally mark to a numeral-token of the same type as x. But then, by Proposition 1.1, a numeral-token consisting of a single tally mark is the result of adding one tally mark to a numeral-token of type $\tau(x)$. Since this is impossible, we conclude that '$|$' does not immediately succeed any instantiated numeral-type. We assume, as usual, that uninstantiated numeral-types will behave the same way.

Proposition 1.7 '$|$' *does not immediately succeed any numeral-type.*

1.3 How Many Types?

Each numeral-type has *at most* one immediate successor. Does each numeral-type have *at least* one immediate successor? I do not know. I would never claim for such a proposition the sort of certainty we normally associate with mathematical results. I am pretty sure, though, that the proposition is not fundamentally incoherent: it seems *logically possible*. In a logically impossible scenario, anything goes: in classical logic, *everything* follows from a contradiction. We, however, are going to suppose that we can distinguish between what would and what would not be the case if every numeral-type had an immediate successor. So it should be worthwhile to explore what things would be like if this were really so. Furthermore, it seems *conceptually possible* that every numeral-type has an immediate successor. That is, this would be compatible with our concept of a numeral-type. So, when we explore what things would be like under these circumstances, we continue to investigate *numeral-types*, not just things we *call* "numeral-types" without any definite idea of what, if anything, we are really discussing.

If $\mathfrak{b}$ is *an* immediate successor of $\mathfrak{a}$, Proposition 1.5 allows us to describe it as *the* (one and only) immediate successor of $\mathfrak{a}$. This justifies some new terminology:

$$|| = S(|)$$
$$||| = S(||)$$
$$|||| = S(|||)$$

and so on. S is the operation or function that takes us from a numeral-type to its immediate successor. In our new notation, Propositions 1.6 looks like this

$$S(\mathfrak{a}) = S(\mathfrak{b}) \implies \mathfrak{a} = \mathfrak{b}$$

while Proposition 1.7 looks like this

$$| \neq S(\mathfrak{a})$$

with '$\mathfrak{a}$' and '$\mathfrak{b}$' understood to be variables that range over all numeral-types (that is, symbols that allow us to make claims about all numeral-types). These two propositions let us prove that there are infinitely many numeral-types (at least in the possible world we are exploring). To get an idea of why this is so, suppose there are only three numeral-types: $|$, $S(|)$, and $S(S(|))$. Each numeral-type has an immediate successor. What, then, is the immediate successor of $S(S(|))$? By Proposition 1.7, $| \neq S(S(S(|)))$. On the other hand, if $S(|) = S(S(S(|)))$, then Proposition 1.6 implies that $| = S(S(|))$, contrary to Proposition 1.7. Finally, if $S(S(|)) = S(S(S(|)))$, then two applications of Proposition 1.6 imply that $| = S(|)$, contrary to Proposition 1.7. So $|$, $S(|)$, and $S(S(|))$ cannot really be the only numeral-types because that would leave $S(S(|))$ without an immediate successor.

I should point out that I am being a little sloppy with my notation. We usually use quotation marks when we name names. $|\,|$ is the number whose name in our system of numeration is '$|\,|$'. When I write

$$|\,|\,| = S(|\,|)$$

this may appear to say that the number $|\,|\,|$ is the result of applying S to the number $|\,|$. The equation

$$\text{'}|\,|\,|\text{'} = S(\text{'}|\,|\text{'})$$

more clearly states that the *numeral-type* '$|\,|\,|$' is the result of applying S to the *numeral-type* '$|\,|$'. A disadvantage of this more careful approach is that writing down all the quotation marks becomes prohibitively tedious after the first few dozen pairs. In self-defence, we adopt the convention that tokens of, say, '$|\,|$' can name '$|\,|$' itself. That is, two tally marks without quotation marks can instantiate the numeral-type '$|\,|$' *and* name it.

If tokens of type $\mathfrak{a}$ name object x, it seems reasonable to say that type $\mathfrak{a}$ is itself a name of x: types name what their tokens name. So, if tokens of a numeral-type are to name that very numeral-type, the numeral-type will name *itself*. Is this a problem? It will turn out that we are exploring a world in which our numeral-types have all the mathematically important properties of the numbers they are supposed to name. So it should not create mathematical difficulties if we let our numeral-types name themselves: they will still be naming things that behave like numbers. This may or may not render our numeral-types ambiguous. Each numeral-type will, indeed, name both a numeral-type (itself) and a number. Furthermore, at the beginning of this chapter, I emphasized the distinction between a name and what the name names. In general, this is an important thing to keep straight. But you might consider whether, in this special case, the distinction is unwarranted. When imagining a situation in which numeral-types have all the mathematically important properties we normally attribute to numbers, it may be convenient to *identify* each numeral-type with the number it names. There will then be no ambiguity: each numeral-type will name exactly one thing. Furthermore, there may be a philosophical benefit. If you understand what numeral-types are, you will know what numbers are. What are numbers? They are numeral-types. Conversely, if you understand what numbers are, you will know what numeral-types are. What are numeral-types? They are numbers. On the other hand, if you already understand what both numeral-types and numbers are and you understand them to be different things, you may not be so happy with my proposal to identify them. I will not insist upon such an identification.

1.4 Recursive Definition

In number theory, we can define primality (the property of being prime) by stating necessary and sufficient conditions for a number to be prime.

> *A natural number is prime if and only if it is greater than one and its only divisors are one and itself.*

We could also define primality by describing a mechanical procedure for determining whether a number is prime. Computer science majors could accomplish this by writing a program. Our inspiration in this section will be definitions of this latter sort. We are going to define operations on numeral-types by offering procedures rather than verbal equivalents. Our first example is a definition of addition.

Definition 1.4

$$a + | = S(a)$$

$$a + S(b) = S(a + b).$$

To get some practice with Definition 1.4, we are going to use it to figure out what $|| + |||$ is. Our approach will not be very subtle: we will figure out what $|| + |$ is; we will use that information to figure out what $|| + ||$ is; then we will use *that* information to figure out what $|| + |||$ is. This is the sort of thing a machine might do, relentlessly advancing one little step at a time with no great intuitive leaps. As for the nuts-and-bolts of applying Definition 1.4, we are, once again, using the variables 'a' and 'b' to make claims about *all* numeral-types: our definition says that certain relationships will hold no matter what a and b are. So we should feel free to replace any occurrences of 'a' or 'b' with any strings of tally marks we wish. Replacing 'a' with '||' in the first clause of the definition, we get:

$$|| + | = S(||) = |||.$$

So now we know what $|| + |$ is. Applying this result and the second clause of the definition, we determine that

$$|| + || = || + S(|) = S(|| + |) = S(|||) = ||||.$$

So now we know what $|| + ||$ is. Repeating this procedure, we determine that

$$|| + ||| = || + S(||) = S(|| + ||) = S(||||) = |||||.$$

We conclude that $||+||| = |||||$. More generally, we have a mechanical procedure for determining what $a + b$ is as long a and b are presented to us in a way that allows us to determine how many tally marks their tokens would have. (If we only know that a and b are God's two favorite numeral-types, we are in trouble.) Just a bit ago, we discussed one way of so presenting them: using their own tokens to name them; for example, writing

$$|||$$

to refer to the type "numeral consisting of three tally marks."

Definition 1.4 is an example of a RECURSIVE DEFINITION. A recursive definition tells us what value an operation yields when we feed it '|' and then tells us how to calculate the output at any successor stage: a calculation in which we apply an operation to the output at the prior stage (Definition 1.4), to the input at the prior stage (Definition 1.5), or to both (Exercise 1.6). Among the operations we are allowed to apply are the successor operation S, the "null operation" that assigns a to each input a, and any operations we have already recursively defined. In Definition 1.4, we apply S to the output at the prior stage. In Definition 1.5, we apply the null operation to the input at the prior stage. A recursive definition allows us to calculate values for arbitrary inputs by starting from '|' and applying a fixed procedure over and over (that is, recursively).

Definition 1.5

$$pred(|) = |$$

$$pred(S(a)) = a.$$

'$pred$' stands for "predecessor." We could think of it as the instruction "erase one tally mark," with the slight glitch that, if you start with just one tally mark, you leave it be. (We are supposing there is no numeral-type whose tokens consist of no tally marks.) So

$$pred(||) = pred(S(|)) = | = pred(|).$$

That is, $pred$ treats $|$ as the "predecessor" of both $|$ and $||$. Otherwise, $pred$ is well-behaved.

$$pred(|||) = ||$$

$$pred(||||) = |||$$

and so on.

Exercise 1.3 *Letting the variable 'a' range over people, define the function* child *as follows.*

$$child(\text{ the father of } a) = a$$

Give an example of an absurd conclusion that follows from this definition. What is there about S that makes our definition of pred legitimate while the definition of child is illegitimate?

Definition 1.6

$$\mathfrak{a} \dot- | = pred(\mathfrak{a})$$

$$\mathfrak{a} \dot- S(\mathfrak{b}) = pred(\mathfrak{a} \dot- \mathfrak{b})$$

Exercise 1.4 *Use Definitions 1.5 and 1.6 to calculate* $|\,|\,|\,|\,| \dot- |\,|\,|$.

Exercise 1.5 *Use Definitions 1.5 and 1.6 to calculate* $| \dot- |\,|\,|$.

Exercise 1.6 *Recursively define the function* $f(\mathfrak{a}) = | + |\,| + |\,|\,| + \cdots + \mathfrak{a}$ *(letting* $f(|) = |$).

1.5 Proof by Induction

Although nothing I have said so far rules out numeral-types "infinitely distant" from '|', I really do mean to rule them out. The idea is that, if $\mathfrak{a}$ is a numeral-type, then

$$\mathfrak{a} = S(S(S(\ldots S(|)\ldots)))$$

where the ellipses '...' represent finitely many applications of the successor operation. Mathematical induction is an attempt to capture this idea. It is also a powerful method of proof.

If '|' has a disease and each numeral-type infects its immediate successor, how many numeral-types will catch the disease? '|\,|\,|\,|' will catch it from '|\,|\,|' who caught it from '|\,|' who caught it from '|'.

$$|\,\rightsquigarrow\,|\,|\,\rightsquigarrow\,|\,|\,|\,\rightsquigarrow\,|\,|\,|\,|$$

'|\,|\,|\,|\,|\,|' will catch it from '|\,|\,|\,|\,|' who caught it from '|\,|\,|\,|'.

$$|\,|\,|\,|\,\rightsquigarrow\,|\,|\,|\,|\,|\,\rightsquigarrow\,|\,|\,|\,|\,|\,|.$$

In fact, every numeral-type will catch the disease because every numeral-type is only finitely many successor steps away from '|' and, so, stands at the end of one of these chains of infection.

This gives us a way to show that every numeral-type has some property. First, show that '|' has the property. Then show that the property is "infectious": that $S(\mathfrak{a})$ has it whenever $\mathfrak{a}$ does. In more standard terminology, the goal is to show that the property is HEREDITARY. If '|' has a hereditary property, it will follow that every numeral-type inherits the property. Here are some examples.

Theorem 1.1 $| + \mathfrak{b} = S(\mathfrak{b})$.

Proof We first confirm that '|' has the desired property. This would mean that

$$| + | = S(|)$$

which is an immediate consequence of Definition 1.4. We next assume the INDUCTIVE HYPOTHESIS that an arbitrary numeral-type has the desired property. Actually, no numeral-type is more or less arbitrary than any other. It is we who are to *treat* the numeral-type as arbitrary by not using any special information about it: by only making inferences that would apply to any other numeral-type. To guarantee that we do not cheat, we will give our "arbitrary numeral-type" a silly name that reveals nothing about it: say, '$\smile$'. Our inductive hypothesis is that

$$| + \smile = S(\smile).$$

We now try to confirm that $\smile$ passes this property on to $S(\smile)$. We want to show that the preceding equation still holds when we replace each '$\smile$' with '$S(\smile)$'. When we perform this substitution on the left side of our inductive hypothesis, we obtain

$$| + S(\smile).$$

When we perform this substitution on the right side of our inductive hypothesis, we obtain

$$S(S(\smile)).$$

So our goal is to verify that

$$| + S(\smile) = S(S(\smile)).$$

We now find equations that take us from the left side of this equation to the right side. By Definition 1.4,

$$| + S(\smile) = S(| + \smile).$$

Applying S to both sides of our inductive hypothesis, we obtain

$$S(| + \smile) = S(S(\smile)).$$

So, as desired,

$$| + S(\smile) = S(S(\smile)).$$

That is, $S(\smile)$ inherits the desired property from $\smile$. Since $\smile$ could be any numeral-type, we have shown that the property is hereditary. So, since '|' has it, every numeral-type has it.

Theorem 1.2 $S(a) + b = S(a + b)$.

Proof Since we are faced with two variables, 'a' and 'b', there are a couple of ways to approach this problem. We could try to show that the formula

$$S(\underline{\quad}) + b = S(\underline{\quad} + b)$$

comes out true no matter how we fill in the blank. Or we could try to show that the formula

$$S(a) + \underline{\quad} = S(a + \underline{\quad})$$

comes out true no matter how we fill in the blank. Let us try the second approach. (If we get stuck, we can go back and try the first approach.) To begin, we want to confirm that we get something true when we put '|' in the blank. That just requires two applications of Definition 1.4:

$$S(a) + | = S(S(a)) = S(a + |).$$

Now we deploy our inductive hypothesis:

$$S(a) + \smile = S(a + \smile).$$

An inductive hypothesis is not always immediately useful. You may have to wander around a bit before you see why it is helpful. Just don't forget about it! Our goal is to show that $S(\smile)$ behaves like $\smile$. That is, we want to confirm that

$$S(a) + S(\smile) = S(a + S(\smile)).$$

As in the preceding proof, we pick one side of this equation and try to transform it into the other side. We will go from left to right, moving from $S(a) + S(\smile)$ to $S(a + S(\smile))$. Definition 1.4 assures us that

$$S(a) + S(\smile) = S(S(a) + \smile).$$

So, by our inductive hypothesis,

$$S(a) + S(\smile) = S(S(a + \smile)).$$

Another application of Definition 1.4 now yields

$$S(a) + S(\smile) = S(a + S(\smile))$$

just as we hoped.

You might find Theorems 1.1 and 1.2 useful in some of the exercises below. In Exercise 1.8, your inductive hypothesis could have the form:

$$(a + \smile) \mathbin{\dot-} \smile = a.$$

Of course, you do not *have* to use the smiley face; but, whatever symbol you use, you are to treat it as if it were a constant term: the name of a numeral-type. The term 'a', on the other hand, is a variable that ranges over all numeral-types. Your inductive hypothesis is a claim about *all* numeral-types a. So, whenever it seems useful, you can feel free to replace 'a' with something else. For example:

$$(S(a) + \smile) \mathbin{\dot-} \smile = S(a).$$

Exercise 1.7

$$a + b = b + a.$$

Exercise 1.8

$$(a + b) \mathbin{\dot-} b = a.$$

Exercise 1.9

$$a \mathbin{\dot-} (b + c) = (a \mathbin{\dot-} b) \mathbin{\dot-} c.$$

Exercise 1.10

$$| \mathbin{\dot-} b = || \mathbin{\dot-} b = |.$$

1.6 A Characteristic Function

After two more exercises, we will consider a new function. When you do these and any subsequent exercises, feel free to use the results of previous exercises.

Exercise 1.11

$$S(a) \mathbin{\dot-} a = |.$$

Exercise 1.12

$$S(a) \mathbin{\dot-} S(b) = a \mathbin{\dot-} b.$$

Now for the new function.

Definition 1.7

$$id(a, b) = |||| \mathbin{\dot-} ((S(a) \mathbin{\dot-} b) + (S(b) \mathbin{\dot-} a)).$$

Exercise 1.13

$$id(a, b) = id(b, a) = id(S(a), S(b)).$$

Since no sum of numeral-types can be | and, hence, $(S(a) \mathbin{\dot-} b) + (S(b) \mathbin{\dot-} a)$ cannot be |, $id(a, b)$ has to be of the form

$$||||\mathbin{\dot-}|| \quad \text{or} \quad ||||\mathbin{\dot-}||| \quad \text{or} \quad ||||\mathbin{\dot-}|||| \quad \text{or} \quad ||||\mathbin{\dot-}||||| \quad \text{or} \quad \ldots$$

So $id(a, b)$ has only two possible values: | and ||. We can express this as an equation

$$pred(id(a, b)) = |.$$

(Note that $pred$ yields | only when applied to | or ||. So this is an indirect way of saying that $id(a, b)$ is | or ||.) The next two exercises will let you verify that this equation is correct.

Exercise 1.14

$$pred(a \mathbin{\dot-} b) = pred(a) \mathbin{\dot-} b.$$

Exercise 1.15

$$||| \mathbin{\dot-} (a + b) = |.$$

Here are two exercises that tell us something about when id yields each of its two possible values.

Exercise 1.16

$$id(a, a) = ||.$$

Exercise 1.17

$$id(a, a + b) = |.$$

A CHARACTERISTIC FUNCTION answers a yes-or-no question about the values we feed it. When we feed a pair of values to id, the question it answers is whether the values are the same. || mean "yes" and | means "no." That is,

$$id(a, b) = \begin{cases} || & \text{if } a = b \\ | & \text{otherwise.} \end{cases}$$

This may seem like a pretty boring function, but it is of interest to logicians because it allows us to define logical expressions such as 'not', 'or', 'only if' and more. This shows that some parts of logic can be captured in elementary arithmetic. Although philosophers, logicians, and the occasional mathematician argue energetically about the proper analysis of various logical expressions, the behavior of the logical operators definable in our theory of numeral-types is well understood and not a source of controversy. We will consider some of these operators in the next section. First, though, we prove an obscure looking fact that will turn out to be useful.

Theorem 1.3 $id(a, b) + id(id(a, b), |) = |||.$

The next two exercises supply an inductive proof of this theorem.

Exercise 1.18
$$id(a, |) + id(id(a, |), |) = |||.$$

Exercise 1.19 *As an inductive hypothesis, assume that*

$$id(a, \smile) + id(id(a, \smile), |) = |||.$$

Your goal is to show that

$$id(a, S(\smile)) + id(id(a, S(\smile)), |) = |||.$$

You might do a second induction to confirm that this really does hold for every a. *You would then want to prove:*

$$id(|, S(\smile)) + id(id(|, S(\smile)), |) = |||$$

and

$$id(S(\heartsuit), S(\smile)) + id(id(S(\heartsuit), S(\smile)), |) = |||.$$

('$\heartsuit$' is another one of our names for an "arbitrary numeral-type." As you reason about $\heartsuit$, you should remember that you are allowed to replace 'a' in our earlier inductive hypothesis with any terms you wish, including terms in which '$\heartsuit$' appears.)

1.7 Some Logic

Back in Sect. 1.3, I assumed that inequalities (such as: '$| \neq S(a)$') were well-understood. As natural as this assumption was, we can now see that it is unnecessary. Each inequality

$$\underline{\quad} \neq \underline{\quad}$$

is equivalent to an equation

$$id(\underline{\quad}, \underline{\quad}) = |.$$

We can always say that some numeral-types are *different* by asserting that some other numeral-types are the *same*. So we could have treated '$\neq$' as a *defined* expression. Let us make a fresh start and do just that.

Definition 1.8 $a \neq b \iff id(a, b) = |.$

Exercise 1.20
$$| \neq S(a).$$

When it seems convenient, we move the negation sign to the front:

$$-(a = b) \iff a \neq b.$$

This convention and Definition 1.8, yield the following equivalences.

$$id(id(a, b), |) = | \iff -(id(a, b) = |) \iff -(-(a = b)).$$

The right-hand formula (the one with the two negation signs) is called a DOUBLE NEGATION. Assessing multiple negations of our equations requires no philosophical or linguistic analysis: it is simply a matter of calculation. The next exercise asks you to evaluate a triple negation.

Exercise 1.21 *Confirm that* $-(-(-(|||| = ||)))$.

Given any two equations, we can use an inequality to assert that at least one of them is true (though it might not be immediately obvious that the following definition does the trick).

Definition 1.9 $a = b \lor c = \mathfrak{d} \iff id(a, b) + id(c, \mathfrak{d}) \neq ||$

$\lor$ is the DISJUNCTION operator. You can read it as "or." If you have studied logic, you have probably seen the TRUTH TABLE for disjunction.

ϕ	ψ	$(\phi \lor \psi)$
T	T	T
T	F	T
F	T	T
F	F	F

The idea is that a disjunction is false if and only if each of its components is false. If I assert "ϕ or ψ," I am asserting that at least one of the alternatives ϕ, ψ is true. So I am wrong if and only if they are both false. Now consider the following table.

$\alpha = \beta$	$\gamma = \delta$	$id(\alpha, \beta)$	$id(\gamma, \delta)$	$id(\alpha, \beta) + id(\gamma, \delta)$	$\alpha = \beta \lor \gamma = \delta$
T	T	||	||	||||	T
T	F	||	|	|||	T
F	T	|	||	|||	T
F	F	|	|	||	F

If I assert "$\alpha = \beta$ or $\gamma = \delta$," I am asserting that $id(\alpha, \beta)$ and $id(\gamma, \delta)$ are not both | and, hence, that $id(\alpha, \beta) + id(\gamma, \delta)$ is not ||. Definition 1.9 guarantees that the disjunction of two equations is false if and only if both equations are false. In the next exercise, you will prove a disjunction central to classical logic. This might be a good time to recall Theorem 1.3.

Exercise 1.22

$$a = b \lor a \neq b.$$

This is a version of the LAW OF EXCLUDED MIDDLE or **LEM**. Each instance of **LEM** offers two alternatives: "Either ϕ or not ϕ." What is *excluded* is some alternative beyond these two. You might check to see whether we used **LEM** in any of our reasoning above. (If so, we might just be reasoning in a circle.)

Exercise 1.23 *Look up Grelling's Paradox and be prepared to say why the following instance of* **LEM** *is questionable:* "The adjective 'heterological' either is or is not heterological."

The following definition will help us introduce another logical operator.

Definition 1.10

$$\sum_{i=|}^{|} f(i) = f(|)$$

$$\sum_{i=|}^{S(a)} f(i) = f(S(a)) + \sum_{i=|}^{a} f(i)$$

$\sum$ is the SUMMATION function. That is,

$$\sum_{i=|}^{a} f(i) = f(|) + f(||) + \cdots + f(a).$$

You can let f be any function you can define using the resources of this chapter. For example, you could let $f(i)$ be $S(i) \dot- i$, as in Exercise 1.24, or $id(i, |||)$, as in Exercise 1.25.

Exercise 1.24

$$\sum_{i=|}^{a} (S(i) \dot- i) = a.$$

Exercise 1.25

$$\sum_{i=|}^{||} id(i, |||) = ||.$$

We can generalize the result of the preceding exercise. Consider any sum of the following form:

$$\sum_{i=|}^{a} id(f(i), b) = id(f(|), b) + id(f(||), b) + \cdots + id(f(a), b).$$

If each of the equations

$$f(|) = \mathfrak{b}, \ f(||) = \mathfrak{b}, \ \ldots \ , f(\mathfrak{a}) = \mathfrak{b}$$

were *false*, then each of the terms

$$id(f(|), \mathfrak{b}), \ id(f(||), \mathfrak{b}), \ \ldots \ , id(f(\mathfrak{a}), \mathfrak{b})$$

would equal | and, hence, their sum would equal $\mathfrak{a}$. Their sum would differ from $\mathfrak{a}$ if and only if some of the equations

$$f(|) = \mathfrak{b}, \ f(||) = \mathfrak{b}, \ \ldots \ , f(\mathfrak{a}) = \mathfrak{b}$$

were true. This leads us to the following definition.

Definition 1.11

$$\exists x \leq \mathfrak{a} \ f(x) = \mathfrak{b} \iff \sum_{i=|}^{\mathfrak{a}} id(f(i), \mathfrak{b}) \neq \mathfrak{a}.$$

Definition 1.11 introduces the BOUNDED EXISTENTIAL QUANTIFIER: $\exists x \leq$ ___. It is to be read as, "There is an x no greater than ___ such that ...." For example, the sentence

$$\exists x \leq ||||| \ S(x) = |||$$

says there is a numeral-type no greater than ||||| whose successor is |||. Here ||||| forms an upper bound for the values of the variable 'x'. If you complain at this point that I have not yet defined "no greater than," I will have to plead guilty. I have defined the expression '$\exists x \leq |||||$' without defining the expression '$\leq$'. That does not make the definition any less legitimate. You could apply the definition without having any independent idea of what '$\leq$' might mean. Informally, though, it might make the definition easier to digest if you think of $\leq$ as the natural ordering of numeral-types in terms of the number of tally marks in their tokens:

$$| \leq || \leq ||| \leq |||| \leq ||||| \leq |||||| \leq \cdots$$

Exercise 1.26
$$\exists x \leq ||| \ x + || = ||||.$$

To indicate how useful our new quantifier is, I will use it to define divisibility and primality. First, though, I need to define multiplication and the CONJUNCTION operator $\wedge$ (to be read as "and").

Definition 1.12
$$\mathfrak{a} \cdot | = \mathfrak{a}$$

$$\mathfrak{a} \cdot S(\mathfrak{b}) = \mathfrak{a} \cdot \mathfrak{b} + \mathfrak{a}.$$

Definition 1.13 $a = b \,\wedge\, c = \eth \iff id(a, b) + id(c, \eth) = ||||$.

Exercise 1.27 *Explain Definition 1.13 in the way we earlier explained Definition 1.9. (You might start by supplying the truth table for conjunction.)*

To express the idea that $\eth$ divides a without remainder, we start with the scheme

$$\exists x \leq a \; f(x) = b$$

from Definition 1.11 and, letting

$$f(x) = x \cdot \eth$$
$$b = a$$

we obtain

$$\exists x \leq a \; x \cdot \eth = a.$$

$\eth$ divides a without remainder if and only if a is the product of $\eth$ and a term x no greater than a. Since we need not look beyond a when searching for the quotient x, a bounded existential quantifier meets our needs perfectly well.

Definition 1.14 $\eth | a \iff \exists x \leq a \; x \cdot \eth = a$.

Definition 1.15 $prime(a) \iff a \neq | \,\wedge\, \neg\exists x \leq pred(a)(x \neq | \,\wedge\, x | a)$.

When testing a for primality, our search for a's divisors does not even carry us as far as a. So we can, once again, make do with a bounded existential quantifier.

Exercise 1.28 *Define the* BOUNDED UNIVERSAL QUANTIFIER $\forall x \leq$ ___ (to be read as: "for all x no greater than ___").

1.8 Π_1^0 Sentences

The PRIMITIVE RECURSIVE functions are those obtainable from the numeral | and the function S by composition and recursive definition.[1] Definitions 1.4, 1.5, 1.6, 1.10, and 1.12 are all recursive. Definition 1.7 introduces the function id by composing S and two recursively defined functions (addition and subtraction). Definitions 1.8, 1.9, 1.11, 1.13, 1.14, and 1.15 just introduce abbreviations for expressions already defined. So all the functions defined in the previous four sections are primitive recursive. Furthermore, all the defined predicates, relations, and logical operators can be

---

[1] I am not going to supply rigorous definitions of composition and recursion. A good source for a more thorough treatment is the WIKIPEDIA article on primitive recursive functions (http://en. wikipedia.org/wiki/Primitive_recursive_function). For an especially meticulous (though perhaps not so readable) presentation, see Curry [3] (http://www.jstor.org/stable/2371522).

replaced by primitive recursive characteristic functions. Indeed, every proposition we can formulate with our defined expressions (and with any expressions similarly defined) will be equivalent to a sentence of the form

$$f(\mathfrak{a}) = |$$

where f is primitive recursive.

Is that obvious? Hardly. You may even even have doubts. One reason for skepticism is that we can easily formulate propositions that have no occurrences of '$\mathfrak{a}$' or any other variable. How can a variable-free formula be equivalent to a generalization of the form $f(\mathfrak{a}) = |$? There are, in fact, all sorts of sneaky ways to arrange for such an equivalence. For example, the formula

$$S(|) \neq |$$

is equivalent to the generalization

$$id(S(| \dot{-} \mathfrak{a}), | \dot{-} \mathfrak{a}) = |$$

since Exercise 1.10 guarantees that $| \dot{-} \mathfrak{a} = |$.

Exercise 1.29 *Write down a sentence of the form $f(\mathfrak{a}) = |$ (f primitive recursive) equivalent to '$| + | = ||$'.*

Another worry is that some of our sentences feature more than one variable. How could a sentence like '$\mathfrak{a} + \mathfrak{b} = \mathfrak{b} + \mathfrak{a}$' be equivalent to a sentence with occurrences of only one variable? The answer is that there are tricks that allow us to code up finite sequences of terms using just one term. We will now consider one such trick. First, we define exponentiation.

Definition 1.16
$$\mathfrak{a}^| = \mathfrak{a}$$
$$\mathfrak{a}^{S(\mathfrak{b})} = \mathfrak{a} \cdot \mathfrak{a}^{\mathfrak{b}}.$$

It would now be helpful to have a characteristic function for divisibility. That is, we would like a function div such that

$$div(\mathfrak{d}, \mathfrak{a}) = \begin{cases} || & \text{if } \mathfrak{d}|\mathfrak{a} \\ | & \text{otherwise.} \end{cases}$$

To confirm that we can define such a function, note that the following are equivalent.

$$\exists x \leq a \; x \cdot \eth = a$$

$$\sum_{i=|}^{a} id(i \cdot \eth, a) \neq a$$

$$id(\sum_{i=|}^{a} id(i \cdot \eth, a), a) = |$$

$$id(id(\sum_{i=|}^{a} id(i \cdot \eth, a), a), |) = ||.$$

So, since $\eth | a$ if and only if $\exists x \leq a \; x \cdot \eth = a$, we can let

$$div(\eth, a) = id(id(\sum_{i=|}^{a} id(i \cdot \eth, a), a), |).$$

Now consider

$$(\sum_{i=|}^{||^{||}} div(||^{i}, ||^{||})) \dot{-} ||^{||}$$

or, to revert to more familiar terminology,

$$(\sum_{i=1}^{2^{2}} div(2^{i}, 2^{2})) \dot{-} 2^{2}.$$

2^1 and 2^2 divide 2^2, while 2^3 and $2^4 (= 2^{(2^2)})$ do not. So the first two terms of our summation are 2, while the last two are 1. That is,

$$(\sum_{i=1}^{2^{2}} div(2^{i}, 2^{2})) \dot{-} 2^{2} = (2 + 2 + 1 + 1) \dot{-} 2^{2} = 2.$$

Similarly,

$$(\sum_{i=1}^{2^{3}} div(2^{i}, 2^{3})) \dot{-} 2^{3} = (2 + 2 + 2 + 1 + 1 + 1 + 1 + 1) \dot{-} 2^{3} = 3.$$

Note, too, that

$$(\sum_{i=1}^{3^{2}} div(3^{i}, 3^{2})) \dot{-} 3^{2} = (2 + 2 + 1 + 1 + 1 + 1 + 1 + 1 + 1) \dot{-} 3^{2} = 2.$$

If a does not divide c, then

$$(\sum_{i=1}^{a^b \cdot c} div(a^i, a^b \cdot c)) \dot{-} a^b \cdot c = b.$$

This motivates the following definition.

Definition 1.17

$$exp(a, b) = (\sum_{i=1}^{b} div(a^i, b)) \dot{-} b.$$

If a divides b, then $exp(a, b)$ is the number of times it does so. For example,

$$exp(2, 432) = exp(2, 2^4 \cdot 3^3) = 4$$

while

$$exp(3, 432) = exp(3, 2^4 \cdot 3^3) = 3.$$

Some may find it helpful to note that if a is even, then $exp(2, a)$ is the exponent of 2 in the prime factorization of a. More generally, we can use exp to extract the exponent from any term in a prime factorization.

How does this help us use a single term to code up a sequence of terms? Suppose we want to express the commutativity of addition using just one variable. We would be looking for a sentence that implies each instance of the generalization

$$a + b = b + a.$$

Try this:

$$exp(2, a) + exp(3, a) = exp(3, a) + exp(2, a).$$

Since this is itself an instance of the generalization we are trying to capture, it does not say more than it should. But does it say *enough*? For example, does it imply that $5 + 4 = 4 + 5$? To see that it does consider the following instance:

$$exp(2, 2592) + exp(3, 2592) = exp(3, 2592) + exp(2, 2592).$$

Note that $2592 = 2^5 \cdot 3^4$. So $exp(2, 2592) = 5$ and $exp(3, 2592) = 4$. This yields the desired conclusion: $5 + 4 = 4 + 5$. There is nothing special about 4 and 5: any two numbers would have worked. So

$$exp(2, a) + exp(3, a) = exp(3, a) + exp(2, a)$$

does express the commutativity of addition using just one variable. Putting everything in the canonical form $f(a) = |$ takes just a little more work. One solution is

$$id(id(exp(|\,|, \mathfrak{a}) + exp(|\,|\,|, \mathfrak{a}), exp(|\,|\,|, \mathfrak{a}) + exp(|\,|, \mathfrak{a})), |) = |$$

which is the double negation of '$exp(|\,|, \mathfrak{a}) + exp(|\,|\,|, \mathfrak{a}) = exp(|\,|\,|, \mathfrak{a}) + exp(|\,|, \mathfrak{a})$'.

Exercise 1.30 *Write down a sentence of the form $f(\mathfrak{a}) = |$ (f primitive recursive) equivalent to* '$\mathfrak{a} \mathbin{\dot-} (\mathfrak{b} + \mathfrak{c}) = (\mathfrak{a} \mathbin{\dot-} \mathfrak{b}) \mathbin{\dot-} \mathfrak{c}$'. *We are helped here by the fundamental theorem of arithmetic: every natural number greater than 1 has a unique prime factorization.*

A sentence is said to be Π^0_1 if it is equivalent to a sentence of the form $f(\mathfrak{a}) = |$ where f is primitive recursive. You might think that Π^0_1 sentences are all quite trivial, but that is wrong. The following two sentences are Π^0_1. (In the second sentence, I write '$|\,| \nmid \mathfrak{a}$' for "$|\,|$ does not divide $\mathfrak{a}$.")

$$\mathfrak{a}^{\partial+|\,|} + \mathfrak{b}^{\partial+|\,|} \neq \mathfrak{c}^{\partial+|\,|}$$

$$|\,| \nmid \mathfrak{a} \ \lor \ \mathfrak{a} = |\,| \ \lor \ \exists x \leq \mathfrak{a} \exists y \leq \mathfrak{a}(prime(x) \land prime(y) \land x + y = \mathfrak{a}).$$

The first is Fermat's Last Theorem; the second is Goldbach's Conjecture (that every even integer greater than two is the sum of two primes). A measure of their depth is the failure of mathematicians as profound as Euler and Gauss to either prove or refute them. Fermat's Last Theorem was proved in 1994. Goldbach's Conjecture remains unresolved.

1.9 Some Philosophy

David Hilbert was convinced that many, though not all, mathematical formulas lack content (*Inhalt* in German): while many formulas might seem to make definite claims about some definite subject matter, they do not really say anything about anything. Although there is room for debate about where exactly Hilbert would draw the line between formulas with and without content, it is reasonably clear that he thought Π^0_1 sentences, interpreted as claims about numerals, have content (or, at least, this is what he ought to have thought given various other things he said).

Hilbert offers some obscure reasons for believing that numerals, particularly our numerals

$$|\quad |\,|\quad |\,|\,|\quad |\,|\,|\,| \quad \cdots$$

are well-suited to be the subject matter of meaningful mathematics. For example, while he seems to concede that we produce tokens of a numeral in various ways, resulting in tokens that differ slightly from one another, he insists that we can recognize in all of them, wherever and whenever they might occur, a single form or shape (*Gestalt*).[2]

Students of Plato might, at this point, recall *Symposium* 211a-d where Diotima discusses beauty itself, "always one in form," not beautiful at one time, ugly at

---

[2] See Hilbert [5], p. 163; English translation in Mancosu [6], p. 202.

another, or beautiful at one place, ugly at another. Indeed, beauty itself *is* the one form shared by all beautiful things, just as a numeral-type is the one form or shape shared by all tokens of that type. Plato was sure that the one form could be better known than any of the many instances: well-founded beliefs about *beauty itself* will be clearer and more reliable than well-founded beliefs about any of the beautiful things in the unstable world of experience. This is not necessarily an otherworldly or unscientific attitude. Note, for example, that physicists are not interested in any electrons *in particular*: they investigate the laws governing all electrons. These very laws set limits to how much information we can acquire about individual electrons. We might say, with Plato, that the laws are "more knowable" than the distinctive features of the individuals they govern.

Someone with Platonic inclinations, might embrace numeral-types, the shapes of numeral-tokens, as the objects of meaningful mathematics. Hilbert's remarks, however, point more in the direction of numeral-tokens. He says, for example, that the objects of meaningful mathematics are present to us in immediate experience (*unmittelbares Erlebnis*).[3] While this is arguably an apt description of our contact with numeral-tokens, it seems not to apply to numeral-types. Our only "experience" of the types is "mediated" by our experience of the tokens: we acquaint ourselves with the types by looking at the tokens. Furthermore, Hilbert says quite clearly that his numerals are not the shapes shared by the instances, they are the instances that share the shapes.[4] Should we conclude, then, that Hilbert considered our numeral-tokens to be the objects of his meaningful mathematics? No, that would be far too hasty. We have barely begun to make such a case. To mention just one important piece of unfinished business, we have not confirmed that Hilbert would accept our neat dichotomy: numeral-type/numeral-token.

It would be best, at this point, to let the professionals work on fine-grained inter-pretations of Hilbert. Without worrying too much about what Hilbert thought, let us consider what *we* ought to think. Let us get some practice using the little bit of philosophical machinery we have discussed in this chapter. For most of the chapter, we have described how numeral-types do or could behave. It has turned out that they do or could behave just like the positive integers (1, 2, 3, . . .). Now we raise a question: would it make sense to insist that our version of arithmetic is also a theory of numeral-tokens: a theory about how numeral-tokens do or could behave? Would it make sense to say that the arithmetical properties we have attributed to numeral-types also apply to numeral-tokens—that numeral-tokens do or could behave just like the positive integers? Let us exercise our philosophical muscles by trying to show that this *would* make sense. I claim only that this enterprise will be educational. Whether we can really fabricate a viable philosophical position along these lines—well, that is something you will have to decide for yourself.

We encounter a problem right away: our theory says that '|' has only one imme-diate successor, yet there are many *tokens* of '||' all of which deserve to be called immediate successors of the many tokens of '|'. As Hilbert himself would be quick to

---

[3] See Hilbert [5], p. 162; Mancosu [6], p. 202.

[4] See Hilbert [5], p. 163, footnote 1; Mancosu [6], p. 214.

remark, mathematicians have a ready-made response to this problem. It is common in mathematics to treat objects as identical when they are really only *equivalent* in some well-defined sense. There are tricks that, in favorable circumstances, allow mathematicians to do this without any logical difficulties.[5] If *sameness-of-shape* is the sort of equivalence relation required for this mathematical sleight of hand, then these tricks would allow us to use the language of shapes when we are really only talking about shaped things.[6] Our talk about "one shape" would then be interpretable as an economical way of talking about the many things so shaped. This would allow us to say that the one and only shape '| |' is the unique immediate successor of the shape '|' even while insisting that the official objects of our theory are numeral-tokens— including the many tokens of '| |'. Or, to put the matter somewhat differently, we could insist that each numeral-token has at most one immediate successor, but when pressed, we would have to acknowledge that we are using the phrase "at most one" in an unusual way.

Having too *few* successors may be a more pressing problem than having too many. If every numeral-token has an immediate successor, there are infinitely many numeral-tokens. But if numeral-tokens are physical, macroscopic objects accessible to us through our senses (as I, for one, have been assuming), it is not so clear why we should believe they are infinite in number. This is reminiscent of our earlier discussion of numeral-*types* in Sect. 1.3. Even in the absence of compelling evidence that there are infinitely many numeral-types, we decided it might be illuminating to explore what things *would* be like if there *were* infinitely many. What is required for such an exploration to be successful? When we set out to explore possible X's, we mean to be discussing the behavior of X's in situations that can be coherently described. If our description would be *incoherent* no matter what we meant to be discussing, we will be trying to pass off as possible something that is *logically* impossible. If our description is incoherent because it is incompatible with our understanding of what it is to be an X, then we will be trying to pass off as possible something that is *conceptually* impossible. It appears that the world of numeral-types, as we have described it so far, is possible in both senses. At least, we have encountered no evidence to the contrary.

Can we say the same about a world inhabited by infinitely many numeral-*tokens*? Is an infinitude of numeral tokens possible both logically and conceptually? As for logical possibility, it is important to remember what our project is right now. We are trying our best to show that the Π_1^0 sentences we would normally treat as true assertions about the positive integers can be reinterpreted as coherent claims about numeral-tokens. In making this case, we should be able to rely entirely on postulates that attribute to numeral-tokens certain abstract, mathematically salient properties: properties that numeral-tokens could share with positive integers. If we really have

---

[5] For an especially clear discussion see Burgess [2], pp. 157–161.

[6] But *is* sameness-of-shape the right sort of relation? Equivalence relations of the sort required are transitive: if a relates to b and b relates to c, then a relates to c. However, the relation *appears-to-be-the-same-shape* is, notoriously, not transitive because imperceptible differences in shape can add up to perceptible ones. Does this consideration not apply to our numeral-tokens?

a fighting chance of showing that numeral-tokens could behave like the positive integers, then our project will probably not require us to emphasize the ways numeral-tokens *differ* from positive integers. So we should be able to reverse the process and reinterpret our claims about numeral-tokens as claims about positive integers.[7] That is, if our theory leads us to say that numeral-tokens have some abstract property, then ordinary arithmetic should lead us to say that the positive integers have that same property. But, if our theory of numeral-tokens were interpretable in ordinary arithmetic, then any absurd consequence of the former theory would be translated into an absurd result of arithmetic. If our theory led us to say that numeral-tokens both have and lack a certain abstract property (to take one example of a palpable absurdity), then ordinary arithmetic would lead us to say that the positive integers both have and lack that property. That is, our theory of numeral-tokens would be inconsistent only if arithmetic itself were inconsistent.[8] So we should worry about the logical possibility of infinitely many numeral-tokens only if we are already worried about the logical possibility of infinitely many positive integers.

Our second question is whether an infinitude of numeral-tokens is conceptually possible. This is a question about our concept of numeral-token and how far we can venture while still discussing things that answer to that concept. Suppose it is indeed part of our concept that numeral-tokens are macroscopic physical objects. Suppose, too, it is part of our concept of physical objects that they are subject to the physical necessities prevailing here in the actual world. That would make it *conceptually* impossible for physical objects to do what is *physically* impossible. So, before we could be confident that an infinitude of numeral-tokens is conceptually possible, we would need to be reassured that it is physically possible. That sounds like a job for the physicists. While we wait to hear from them, we should note that a favorable verdict from the physicists might still leave us with our philosophical question about conceptual possibility: when we discuss imaginary scenarios in which there are infinitely many objects that we *call* "numeral-tokens," are we really discussing numeral-tokens? Even if such scenarios are physically possible, they might conflict in some other way with our concept of numeral-tokens or with the concept we decide we *ought* to have after we think about it a while. Even after getting help from the physicists, we may need to consider very carefully what we mean by a numeral-token before we can make progress.[9] Though we will not pursue the matter further here, I

---

[7] A case of the sort I have in mind is discussed in Chap. 3: when we craft a set theory just strong enough to supply an interpretation of arithmetic, it turns out that arithmetic supplies an interpretation of our set theory. Positive integers are exceptionally powerful devices for coding up all sorts of information—information that may or may not have obvious connections to arithmetic. (We got a taste of this in the previous section when we used individual numbers to code ordered pairs of numbers. There will be more such examples in Chaps. 2 and 3.) It would be surprising if the Π_1^0 part of arithmetic were too weak to code up the mathematically salient claims about numeral-tokens that we need for our reinterpretation of Π_1^0 sentences. Granted, interpretability does not always work both ways. There are theories that interpret arithmetic but are not interpretable in arithmetic. Our project, however, does not seem to require us to formulate such a theory.

[8] Section 5.4 of Chap. 5 may help you see how this works.

[9] To take an extreme example of the sort of thing I have in mind, suppose we try to imagine what it would be like for an elephant to be a pencil. Will we then be thinking about elephants and pencils

do hope *you* will continue to think about it. One of the purposes of this philosophical exercise was to inspire you to help the rest of us think about these issues more clearly.

Some of the above considerations seem not to apply to numeral-types because, at least as I understand them, they are not physical objects subject to physical laws. (If you think they are physical objects, you should be prepared to face the question of *where* they are. Or are they, perhaps, special physical objects that have no spatial location?) Even if you accept that numeral-types are themselves non-physical, you might nonetheless think it conceptually impossible for a numeral-type to exist without having physical tokens inhabiting our universe. (Your slogan might be: no token-less types.) You would then have to concede that the above observations about physical possibility do apply to numeral-types, since any conceptually possible scenario in which there are infinitely many types would be one in which there are infinitely many tokens. But is there really a good reason to believe that numeral-types are subject to such a constraint?

Hilbert's assistant and collaborator PAUL BERNAYS (1888–1977) did not consider either numeral-tokens or numeral-types to be the primary objects of meaningful mathematics. Mathematicians, he says, study *structures*. "Figures" such as

$$| \quad || \quad ||| \quad ||||$$

are of mathematical interest because they exemplify an important structure: the structure formed by the first four ordinal numbers. The figures

$$0 \quad 0' \quad 0'' \quad 0'''$$

exemplify the same structure. The shared structure

$$first \quad second \quad third \quad fourth$$

is the real object of mathematical inquiry. An indication of this is the mathematician's disinterest in the "inessential" differences between the two sequences of figures. The figures '|' and '0' are certainly different, but here they play the same mathematical role. They represent the same position in the sequence of ordinals: the position *first*. It is characteristic of mathematical thought to ignore the differences and focus on the figures' role as representatives of a particular position in an abstract structure.[10]

---

or about things we just *call* "elephants" and "pencils"? For a real-life example, consider what Geraldine Ferraro said when Barack Obama became the front-runner in the race for the Democratic presidential nomination. "If Obama was a white man, he would not be in this position. And if he was a woman of any color, he would not be in this position. He happens to be very lucky to be who he is. And the country is caught up in the concept." Can we really think clearly about a situation in which Barack Obama (not someone *like* Obama, but Obama himself) is, say, a Polynesian woman? As our imaginary excursions become more and more fanciful, could we not reach a point where we are not really talking about Obama?

[10] See Bernays [1], pp. 338–341; English translation in Mancosu [6], pp. 243–245.

We have been considering whether we can supply Π_1^0 sentences with genuine content by interpreting them as claims about concrete objects of sensory experience: that is, numeral-tokens. Bernays, as we just saw, considered this approach strangely out of step with the characteristically mathematical way of thinking. KURT GÖDEL (1906–1978) offers another reservation. When we say that Π_1^0 sentences, interpreted in a certain way, have content, does that claim *itself* have content? In our role as cheerleaders for numeral-tokens, we might say "yes" because we hope to interpret this assertion about Π_1^0 sentences as a claim about concrete objects of sensory experience: namely, Π_1^0 sentence-*tokens*. However, if asked to explain what makes a sentence Π_1^0, we would have to invoke "the general principle of primitive recursive definition" which "contains the abstract concept of function."[11] Since, according to Gödel, functions are not concrete objects of sensory experience, either (1) statements about abstract objects can have content or (2) we have not really managed to say which sentences have content. In the first case, we have no reason to *insist* on a concrete interpretation of Π_1^0 sentences (though we might still consider this an option worth pursuing). In the second case, when we appear to be describing a distinctive philosophical or methodological position, we are actually saying nothing.

1.10 Solutions of Odd-Numbered Exercises

1.1 Suppose $\sigma(xy)$. Pick x', y' such that $\tau(x') = \tau(x)$, $\tau(y') = \tau(y)$, $\upsilon(x'y')$. Now suppose $\sigma(zy)$. Pick z', y'' such that $\tau(z') = \tau(z)$, $\tau(y'') = \tau(y)$, $\upsilon(z'y'')$. Then $\tau(y') = \tau(y'')$ and, hence, by Proposition 1.2, $\tau(x') = \tau(z')$. So $\tau(x) = \tau(z)$. Now go the other direction. Suppose $\tau(x) = \tau(z)$, still assuming that $\sigma(xy)$. Then $\tau(x') = \tau(z)$ and, hence, a numeral-token of the same type as y (namely y') is the result of adding one tally mark to a numeral-token of the same type as z (namely x'). That is, $\sigma(zy)$.

1.3 Barack Obama has two children: Malia and Sasha. According to our definition:

$$\text{child(the father of Malia)} = \text{Malia}$$
$$\text{child(the father of Sasha)} = \text{Sasha.}$$

Furthermore:

$$\text{the father of Malia} = \text{the father of Sasha.}$$

So:

$$\text{Malia} = \text{child(the father of Malia)} = \text{child(the father of Sasha)} = \text{Sasha.}$$

The problem is that children with the same father can be distinct, whereas numeral-types with the same successor cannot.

---

[11] See Gödel [4], p. 272, footnote **b**.

1.5

$$|\dot{-}||| = |\dot{-}S(S(|))$$
$$= pred(|\dot{-}S(|))$$
$$= pred(pred(|\dot{-}|))$$
$$= pred(pred(pred(|)))$$
$$= pred(pred(|))$$
$$= pred(|)$$
$$= |.$$

1.7 Definition 1.4 and Theorem 1.1 confirm that the equation is true when $a = |$. Our inductive hypothesis is that

$$\smile + b = b + \smile.$$

By Definition 1.4, Theorem 1.2, and our inductive hypothesis,

$$S(\smile) + b = S(\smile + b) = S(b + \smile) = b + S(\smile).$$

So the desired property passes from $\smile$ to $S(\smile)$.

1.9 By Definition 1.6,

$$a \dot{-} (b + |) = a \dot{-} S(b) = pred(a \dot{-} b) = (a \dot{-} b) \dot{-} |.$$

So the equation is true when $c = |$. Our inductive hypothesis is that

$$a \dot{-} (b + \smile) = (a \dot{-} b) \dot{-} \smile.$$

By Definitions 1.4 and 1.6,

$$a \dot{-} (b + S(\smile)) = a \dot{-} S(b + \smile) = pred(a \dot{-} (b + \smile)).$$

By Definition 1.6 and our inductive hypothesis,

$$pred(a \dot{-} (b + \smile)) = pred((a \dot{-} b) \dot{-} \smile) = (a \dot{-} b) \dot{-} S(\smile).$$

So

$$a \dot{-} (b + S(\smile)) = (a \dot{-} b) \dot{-} S(\smile).$$

That is, the desired property passes from $\smile$ to $S(\smile)$.

1.11 By Definition 1.4 and Exercises 1.7 and 1.8,

$$S(a) \dot- a = (a + |) \dot- a = (| + a) \dot- a = |.$$

1.13 It follows from Definition 1.7 and Exercise 1.7 that $id(a, b) = id(b, a)$. By Exercise 1.12,

$$|||| \dot- ((S(a) \dot- b) + (S(b) \dot- a)) = |||| \dot- ((S(S(a)) \dot- S(b)) + (S(S(b)) \dot- S(a))).$$

1.15 Our proof is by induction. Note, first, that Definitions 1.5 and 1.6 and Exercises 1.7, 1.9, and 1.10 give us:

$$||| \dot- (a + |) = ||| \dot- (| + a) = (||| \dot- |) \dot- a = || \dot- a = |.$$

By Definitions 1.4, 1.5, and 1.6 and an inductive hypothesis,

$$||| \dot- (a + S(\smile)) = ||| \dot- S(a + \smile) = pred(||| \dot- (a + \smile)) = pred(|) = |.$$

1.17 By Exercises 1.9, 1.10, and 1.11,

$$S(a) \dot- (a + b) = (S(a) \dot- a) \dot- b = | \dot- b = |.$$

By Definition 1.4 and Exercises 1.7 and 1.8,

$$(S(a + b) \dot- a) = (a + S(b)) \dot- a = (S(b) + a) \dot- a = S(b).$$

So Definitions 1.4, 1.6, and 1.7 and Exercises 1.9, 1.10, and 1.15 let us reason as follows

$$\begin{aligned}
id(a, a + b) &= |||| \dot- (| + S(b)) \\
&= (|||| \dot- |) \dot- S(b) \\
&= ||| \dot- S(b) \\
&= ||| \dot- (b + |) \\
&= |.
\end{aligned}$$

1.19 As an inductive hypothesis, we assume that

$$id(a, \smile) + id(id(a, \smile), |) = |||.$$

We want to show that

$$id(a, S(\smile)) + id(id(a, S(\smile)), |) = |||.$$

To that end, we begin a second induction (this time on a). Exercises 1.13 and 1.18 guarantee that

$$id(|, S(\smile)) + id(id(|, S(\smile)), |) = |||.$$

Exercise 1.13 and our earlier inductive hypothesis let us reason as follows

$$id(S(\heartsuit), S(\smile)) + id(id(S(\heartsuit), S(\smile)), |) = id(\heartsuit, \smile) + id(id(\heartsuit, \smile), |) = |||.$$

This completes our induction on $\mathfrak{a}$ (an odd one without its own inductive hypothesis). So we have confirmed that

$$id(\mathfrak{a}, S(\smile)) + id(id(\mathfrak{a}, S(\smile)), |) = |||.$$

This completes our first induction.

1.21 We can use Definition 1.4 to confirm that $|||| = || + ||$. So, by Exercises 1.13 and 1.17, $id(||||, ||) = |$ and, hence, by Exercise 1.16, $id(id(||||, ||), |) = ||$. Definition 1.4 implies that $|| = | + |$. So, by Exercises 1.13 and 1.17, $id(id(id(||||, ||), |), |) = |$. That is, $-(-(-(|||| = ||)))$.

1.23 An adjective is heterological if and only if it does not apply to itself. For example, the adjective 'monosyllabic' is heterological because it is not monosyllabic. What about the adjective 'heterological'? If it is heterological, then it applies to itself and, hence, is not heterological. If it is not heterological, then it does not apply to itself and, hence, is heterological. So if it either is or is not heterological, then it both is and is not heterological. That is, the instance of **LEM** we are considering yields an outright contradiction.

1.25

$$\sum_{i=|}^{||} id(i, |||) = id(|, |||) + id(||, |||)$$
$$= | + |$$
$$= ||.$$

1.27 Here is the truth table for conjunction.

ϕ	ψ	$(\phi \wedge \psi)$
T	T	T
T	F	F
F	T	F
F	F	F

The idea is that a conjunction is true if and only if both its components are true. If I assert "ϕ and ψ," I am asserting that ϕ and ψ are both true. Now consider the following table.

If I assert "$\alpha = \beta$ and $\gamma = \delta$," I am asserting that $id(\alpha, \beta)$ and $id(\gamma, \delta)$ are both $||$ and, hence, that $id(\alpha, \beta) + id(\gamma, \delta)$ is $||||$. Definition 1.13 guarantees that the conjunction of two equations is true if and only if both equations are true.

$\alpha = \beta$	$\gamma = \delta$	$id(\alpha, \beta)$	$id(\gamma, \delta)$	$id(\alpha, \beta) + id(\gamma, \delta)$	$\alpha = \beta \wedge \gamma = \delta$
T	T	||	||	||||	T
T	F	||			|||	F
F	T			||	|||	F
F	F					||	F

1.29 Here is one example.

$$id(id(| + |, ||), | \dot{-} \mathfrak{a}) = |.$$

References

1. Bernays, P. (1930–1931). Die Philosophie der Mathematik und die Hilbertische Beweistheorie. *Blätter für deutsche Philosophie, 4*, 326–367.
2. Burgess, J. P. (2005). *Fixing frege*. Princeton NJ: Princeton University Press.
3. Curry, H. B. (1941). A formalization of recursive arithmetic. *American Journal of Mathematics, 63*, 263–282.
4. Gödel, K. (1990). *Collected works* (Vol. 2). New York: Oxford University Press.
5. Hilbert, D. (1922). Neubegründung der Mathematik. *Abhandlungen aus dem Mathematischen Seminar der Hamburgischen Universität, 1*, 157–177.
6. Mancosu, P. (Ed.). (1998). *From Brouwer to Hilbert*. New York: Oxford University Press.

Chapter 2
Peano Arithmetic, Incompleteness

2.1 The language of PA

We have seen that some profound claims about the natural numbers are Π_1^0. On the other hand, the vocabulary of number theory allows us to formulate sentences that are not Π_1^0. Hilbert would say that at least some of these sentences lack content. We are going to consider a theory replete with such sentences. The theory is Peano Arithmetic (PA), a formalized version of elementary number theory.[1] A formal theory needs a formal language. The language of PA has the following vocabulary.

1. Two CONNECTIVES: '$-$' ("not"), '$\rightarrow$' ("if ...then").
2. A QUANTIFIER: '$\forall$' ("for all").
3. The IDENTITY symbol: '$=$'.
4. Three FUNCTION symbols: 'S' ("successor"), '$+$' ("plus"), '$\cdot$' ("times").
5. One PROPER NAME: '0' ("zero").
6. Infinitely many VARIABLES: 'w', 'x', 'y', 'z', 'w_1', 'x_1', 'y_1', 'z_1', ...
7. Two PARENTHESES: '$($', '$)$'.

In everyday English, certain expressions refer to individual things or can so refer in appropriate contexts. Examples include proper names ('Kurt Gödel'), pronouns ('him'), and descriptions ('the second son of Marianne Gödel'). The TERMS of PA are the expressions that play this role in our formal language. We define them recursively as follows.

8. '0' is a term.
9. Every variable is a term.
10. If α and β are terms, then so are $\ulcorner S\alpha \urcorner$, $\ulcorner(\alpha + \beta)\urcorner$, and $\ulcorner(\alpha \cdot \beta)\urcorner$.

In everyday English, declarative sentences declare that something is the case ('Kurt Gödel was smart') or *could*, if placed in appropriate contexts, declare that

---

[1] Although the 'P' in 'PA' commemorates the mathematician GIUSEPPE PEANO (1858–1932), another mathematician, RICHARD DEDEKIND (1831–1916), deserves much of the credit. For some of the history, see Wang [11] (http://www.jstor.org/stable/2964176).

S. Pollard, *A Mathematical Prelude to the Philosophy of Mathematics*,
DOI: 10.1007/978-3-319-05816-0_2,
© Springer International Publishing Switzerland 2014

something is the case ('He was smart'). The FORMULAS of PA are the expressions that play this role in our formal language. We define them recursively as follows.

11. If α and β are terms, then $\ulcorner \alpha = \beta \urcorner$ is a formula.
12. If ϕ and ψ are formulas, then so are $\ulcorner -\phi \urcorner$ and $\ulcorner (\phi \rightarrow \psi) \urcorner$.
13. If α is a variable and ϕ is a formula, then $\ulcorner \forall \alpha \ \phi \urcorner$ is a formula.

Symbols are said to OCCUR in terms and formulas. For example, the function symbol '+' occurs in the term '$((x + y) + z)$'. In fact, it occurs twice: it has two OCCURRENCES. The left-hand parenthesis '(' also occurs twice, as does the right-hand parenthesis ')'. The variables 'x', 'y', and 'z' each occur once.

The symbols '$\ulcorner$' and '$\urcorner$' are known as CORNER QUOTES.[2] If α is a term, then $\ulcorner S\alpha \urcorner$ is the term consisting of an occurrence of the function symbol 'S' followed by an occurrence of the term α. If α is '0', then $\ulcorner S\alpha \urcorner$ is '$S0$'. Other corner quotations are to be understood similarly. Note that ordinary single quotation marks would not do what we intend here. For example, '$S\alpha$' is not a term of PA even if α is. This is because 'α' is not a term of PA. 'α' is an expression we use to discuss the terms of PA. To take a less technical example: 'David Hilbert' is an expression we use to designate a certain mathematician. To say that 'David Hilbert' was a mathematician is to say that David Hilbert's name was a mathematician. As far as we know, no mathematician has ever been David Hilbert's name (Such an arrangement is unlikely to be practical).

Exercise 2.1 *Suppose α is a PA-term. Is $\ulcorner \alpha S \urcorner$ a PA-term? Is $\ulcorner S\alpha \urcorner$ a PA-term? Does α occur in $\ulcorner S\alpha \urcorner$? Does α occur in '$S\alpha$'? Does 'α' occur in $\ulcorner S\alpha \urcorner$? Does '$\alpha$' occur in '$S\alpha$'?*

The SUB- FORMULAS of a formula ψ are ψ itself and any formulas that occur in ψ. For example, the formula

$$\forall x \forall y \ (x + Sy) = ((x + y) + S0)$$

has three sub-formulas. They are

$$(x + Sy) = ((x + y) + S0),$$

$$\forall y \ (x + Sy) = ((x + y) + S0),$$

and the whole formula itself. Any occurrence of a variable α within a sub-formula of ψ of the form $\ulcorner \forall \alpha \ \phi \urcorner$ is BOUND in ψ. All other occurrences are FREE in ψ. There are three bound occurrences of 'x' in

$$\forall x \forall y \ (x + Sy) = ((x + y) + S0).$$

---

[2] See Quine [9, pp. 33–37]. I will soon get sloppy and stop using the corner quotes (Absolute rigor does get a bit tedious). I thought, however, that you should know about them.

There are two free occurrences of 'x' in

$$\forall y \, (x + Sy) = ((x + y) + S0).$$

A free variable is like a pronoun without an antecedent. The expressions 'He is short' and '$x = 0$', outside of any appropriate context, make no definite claims and, so, are neither true nor false. (Who is *he* and what is x?)

Exercise 2.2 *Identify the bound occurrences of 'x', 'y', and 'z' in the following formula.*

$$\forall y \forall z (-\forall x - (z = Sy \rightarrow z = x) \rightarrow (x + y) = S(x \cdot SS0))$$

In our formal language, we understand SENTENCES to be formulas with no free occurrences of variables. This is a departure from everyday usage where 'He is short' counts as a declarative sentence even in contexts where the pronoun 'He' fails to refer to anyone and, so, is behaving like a free variable. Each sentence of PA makes a definite claim as soon as we supply PA with an interpretation (See Sect. 2.3 below).

2.2 The Axioms of PA

We will now use our formal language to describe the natural numbers $(0, 1, 2, 3, \ldots)$. Actually, we will only describe how the natural numbers are related to one another: we will only describe the *structure* formed by the natural numbers or (speaking more circumspectly) the structure they *would* form if they were to exist. We do this by accepting certain sentences, the AXIOMS, without proof. We show that a sentence is a THEOREM by showing that it follows from the axioms. PA has two axioms characterizing the successor operator. The first says that 0 is not the immediate successor of any natural number ("Given any natural number x, it is not the case that 0 is identical to the immediate successor of x"). The second says that natural numbers with the same immediate successor are the same ("Given any natural numbers x and y, if the immediate successor of x is identical to the immediate successor of y, then x is identical to y").

S1 $\forall x - 0 = Sx$
S2 $\forall x \forall y (Sx = Sy \rightarrow x = y)$

PA has two axioms giving the recursive definition of addition and another two offering the recursive definition of multiplication.

A1 $\forall y \, (y + 0) = y$
A2 $\forall x \forall y \, (y + Sx) = S(y + x)$

M1 $\forall y \ (y \cdot 0) = 0$
M2 $\forall x \forall y \ (y \cdot Sx) = ((y \cdot x) + y)$

Finally, **PA** has infinitely many induction axioms, one for each formula with free occurrences of 'x' and 'y'. That is, if $\phi(x, y)$ is a formula of **PA** with free occurrences of 'x' and 'y', but no free occurrences of any other variable, if $\phi(0, y)$ is the result of replacing each free occurrence of 'x' in $\phi(x, y)$ with an occurrence of '0', and if $\phi(Sx, y)$ is the result of replacing each free occurrence of 'x' in $\phi(x, y)$ with an occurrence of 'Sx', then

$$\ulcorner \forall y(\phi(0, y) \to (\forall x(\phi(x, y) \to \phi(Sx, y)) \to \forall x \ \phi(x, y)))\urcorner$$

is an axiom of **PA**.

An example may clarify the connection between our induction axioms and the proofs by induction we did in Chap. 1. We can let $\phi(x, y)$ be any **PA**-formula with free occurrences only of 'x' and 'y'. If we let $\phi(x, y)$ be '$(y + x) = x$', we obtain the axiom

$$\forall y((y + 0) = 0 \to (\forall x((y + x) = x \to (y + Sx) = Sx) \to \forall x \ (y + x) = x))$$

This says the formula

$$((y + 0) = 0 \to (\forall x((y + x) = x \to (y + Sx) = Sx) \to \forall x \ (y + x) = x))$$

will be true no matter what y is. So, in particular, we are free to replace each free occurrence of 'y' with an occurrence of '0':

$$((0 + 0) = 0 \to (\forall x((0 + x) = x \to (0 + Sx) = Sx) \to \forall x \ (0 + x) = x))$$

This sentence says that you can show

$$\forall x \ (0 + x) = x$$

by first showing

$$(0 + 0) = 0$$

(to get through the first '$\to$') and then showing

$$\forall x((0 + x) = x \to (0 + Sx) = Sx)$$

(to get through the last '$\to$'). The goal here is to show that each natural number x is identical to $0 + x$. The first step is to show that 0 has this property: that 0 is identical to $0 + 0$. The next step is to show that this property is hereditary: that each natural number passes it on to its successor. You might think of the left-hand part of the

formula
$$((0 + x) = x \rightarrow (0 + Sx) = Sx)$$

as an inductive hypothesis, while the right-hand part is the result to be obtained from that hypothesis.

Exercise 2.3 *Using A1, A2, and the induction axiom we have been discussing, prove informally that* $\forall x \ (0 + x) = x$. *(I say "informally" because I have not introduced a formal deductive system of the sort you may have studied in a logic course. You still need to make good inferences when you do this exercise, but you can just go ahead and make those inferences without citing any explicit inference rules).*

2.3 Incompleteness 1: Compactness

As I mentioned above, a theorem of PA is a sentence in the language of PA that follows from the axioms of PA. A PROOF in PA is a demonstration that a PA-sentence does follow from the PA-axioms. A FORMAL LOGIC for PA will include a definition of 'proof' precise enough to yield a mechanical procedure for telling whether something is a proof. The logic of PA, CLASSICAL FIRST- ORDER LOGIC, was first formalized by GOTTLOB FREGE (1848–1925).[3] There are a variety of formalizations equivalent to Frege's. Assume we have picked one and suppose Γ is a set of sentences.

Definition 2.1 ψ is DERIVABLE from Γ if and only if there is a formal proof whose conclusion is ψ and whose premises are members of Γ. If ψ is derivable from Γ, we write: $\Gamma \vdash \psi$. If ψ is derivable from axioms of PA, we write: PA $\vdash \psi$.

We could have characterized the axioms and theorems of PA without offering any hints about what those axioms and theorems *say*. This leaves us free to supply various readings or interpretations. In an INTERPRETATION of PA, we (1) specify the range of our bound variables, (2) assign an object from that range to '0', and (3) assign operations defined on that range to each of 'S', '+', and '·'. If we are not feeling too adventurous we might (1) let our bound variables range over the natural numbers (so that we read '$\forall x$' as "for all natural numbers x"), (2) let '0' be our name for zero, and (3) assign the operations of immediate succession, addition, and multiplication to 'S', '+', and '·'. This, after all, is the *intended* interpretation of PA. You will perhaps agree that all the axioms of PA come out true when so interpreted. An interpretation that makes all the axioms of PA true is said to be a MODEL of PA.

Here is an alternative interpretation. Let our bound variables range over zero and all the negative integers (so that we read '$\forall x$' as "for all non-positive integers x"). Let '0' be our name for zero. Assign the operation "minus one" to 'S' and the operation of addition to '+'. Read '$x \cdot y$' as "x times y times negative one". You might take a

---

[3] See Frege [2]; English translation in van Heijenoort [10, pp. 5–82]. If you have had a logic course, you have probably worked with a close cousin of Frege's system.

minute to confirm that all the axioms of PA come out true when interpreted in this way (So this is a model of PA).

Exercise 2.4 *Let our bound variables range over all the even natural numbers (including zero). Complete this interpretation in a way that makes the first six axioms of PA true.*[4]

I already noted that a proof is a demonstration that a sentence (the conclusion) follows from some sentences (the premises). I did not, however, define "follows from". That was probably fine: you probably already understood this sort of "following" well enough to make sense of the preceding discussion. Insisting on a definition of *everything* is a sure-fire way to block intellectual progress. Sometimes, however, a mathematical definition of a notion that is already well understood is a way of drawing that notion into the mathematical realm: making it the object of productive mathematical inquiry. That is the goal of the following definition.

Definition 2.2 ψ FOLLOWS FROM Γ if and only if it is logically impossible for an interpretation to make all the members of Γ true and ψ false. If ψ follows from Γ, we write: $\Gamma \vDash \psi$. If ψ follows from axioms of PA, we write: PA $\vDash \psi$.

OK, we still have work to do: we do not yet have a mathematical definition of logical impossibility. Supplying such a definition is one of the most important jobs of set theory (You might think of set theory as an inventory of logically possible structures). We will leave a serious exposition of set theory for later chapters. For now, we will try to make do with a less systematic understanding of logical impossibility. Even without a full-blown theory, we have a good idea of how to establish that a situation is logically impossible: assume that the situation is real and derive a contradiction from that assumption.

There is another bit of work we will leave undone: we will not offer a mathematically precise account of what our logical symbols mean. This *can* be done (You may have already seen it done in a logic course. You saw a bit of it done in Chap. 1). We will just not be doing it here. If, in what follows, you need to figure out whether it is logically possible for an interpretation to make certain sentences true or false, you will have to draw on your rough-and-ready understanding of those sentences. I hope it is clear, for example, that no interpretation can make '0 = 0' false since that would require that the object assigned to '0' be distinct from the object assigned to '0'.

It turned out to be mathematically fruitful to give a mathematical definition of the "follows from" relation. In his doctoral dissertation of 1929, for example, Kurt Gödel proved that Frege's formalization of first-order logic is COMPLETE: any conclusion that follows from a set of premises is derivable from that set.[5]

Theorem 2.1 $\Gamma \vDash \psi$ *only if* $\Gamma \vdash \psi$.

---

[4] More modestly, we want an interpretation that *would* make the first six axioms true if there *were* such things as the natural numbers. This sort of conditional claim is what I will generally intend when I talk about an interpretation making certain sentences true.

[5] Gödel's dissertation is reprinted, with an English translation, in Gödel [7, pp. 60–101].

Frege's formalization of first-order logic is also SOUND: any conclusion that is derivable from a set of premises follows from that set.

Theorem 2.2 $\Gamma \vdash \psi$ *only if* $\Gamma \vDash \psi$.

Exercise 2.5 *Formal proofs are all finite: they consist of finitely many lines featuring only finitely many premises. Use this fact, and the preceding two theorems, to prove Gödel's* COMPACTNESS THEOREM: *if a formula follows from* Γ, *it follows from a finite subset of* Γ.

Exercise 2.6 *Suppose it is logically possible for* PA *to have a model. Show that* PA $\nvdash 0 \neq 0$.

Exercise 2.7 *Suppose* PA $\nvdash 0 \neq 0$. *Show that it is logically possible for* PA *to have a model.*

We are going to see that PA suffers from a particular form of incompleteness. First, though, we need to reflect on the relationship between numbers and numerals. The NUMERALS of PA are '0' and any terms consisting of an occurrence of '0' preceded by finitely many occurrences of 'S'. That is:

$$\text{'0', 'S0', 'SS0', 'SSS0', } \ldots$$

We understand the natural numbers to be 0 and everything obtainable from 0 by finitely many applications of the successor operation S. So it follows from our conception of the natural numbers that each of them is named by a numeral of PA when these numerals are interpreted in the standard way (with '0' naming 0 and 'S' expressing the successor operation). For example, the number obtained from 0 by 15 application of S is named by the numeral consisting of an occurrence of '0' preceded by 15 occurrences of 'S'. The STANDARD numbers are 0 and all the numbers obtainable from 0 by finitely many applications of S: that is, exactly the numbers named by our numerals. It is part of our concept of the natural numbers that each of them is standard. So, if PA completely captures our concept of the natural numbers, the axioms of PA will rule out non-standard numbers: that is, it will be logically impossible for an interpretation to make all the axioms of PA true while including in the range of PA's bound variables an object that is not named by any numeral of PA.

At this point it is useful to introduce two new symbols: the proper name 'c' and the inequality symbol '$\neq$'. We use '$\neq$' to make our claims about non-identity a bit more readable:

$$\alpha \neq \beta \iff -\alpha = \beta.$$

You might think of 'c' as the name of a natural number (though I have not said which one). Now consider the following sequence of sentences.

$$c \neq 0, \quad c \neq S0, \quad c \neq SS0, \quad c \neq SSS0, \quad \ldots$$

Let C be the set of all these sentences. Each member of C says something compatible with our conception of the natural numbers. For example,

$$c \neq SSSSS0$$

makes the innocent claim that a certain (so far unidentified) natural number is distinct from five. Given any model of PA, we could make the above sentence true by assigning to 'c' an object in the range of our bound variables. Just let c be anything other than the object named by '$SSSSS0$'.

Exercise 2.8 *Suppose C^* is a proper subset of C. That is, every member of C^* is a member of C, but not every member of C is a member of C^*. Show the following: given any model of PA, we can make all the members of C^* true by assigning to 'c' an object in the range of our bound variables. You may assume the following: If α and β are PA-numerals and if a model of PA makes the equation $\ulcorner \alpha = \beta \urcorner$ true, then $\alpha = \beta$. (Can you see why this is so?)*

Let PA $\cup$ C be the set consisting of the axioms of PA and the members of C. An interpretation of PA $\cup$ C is an interpretation of PA that, in addition, assigns to 'c' an object in the range of PA's bound variables. We now adopt a premise for a conditional proof: that is, an argument intended to establish an "if …then" statement. We need not believe that this premise is true. Our project is to see what follows from it.

Premise (for conditional proof):   PA completely captures our concept of the natural numbers and, so, rules out non-standard numbers.

This would mean that a model of PA $\cup$ C in which c behaves like a non-standard number is logically impossible. On the other hand, if we make all the members of C true, c will *have* to be non-standard, since c will be distinct from every number named by a numeral of PA. So our premise implies that it is logically impossible for an interpretation to make all the members of PA $\cup$ C true. This means that

$$\text{PA} \cup C \vDash \psi$$

no matter what ψ is. (Check the definition of 'follows from' to confirm this.) For example,

$$\text{PA} \cup C \vDash 0 \neq 0.$$

So, by the Compactness Theorem, '$0 \neq 0$' follows from *finitely many* members of PA $\cup$ C. We may suppose, then, that

$$\text{PA}^* \cup C^* \vDash 0 \neq 0$$

where PA* is a finite set of PA-axioms and C^* is a finite subset of C. Any interpretation that makes all the members of PA true will make all the members of PA* true and can easily be extended to make the finitely many members of C^* true (as you showed in the last exercise). Such an interpretation would have to make '$0 \neq 0$' true since '$0 \neq 0$' follows from $PA^* \cup C^*$. But that is logically impossible. (It is logically impossible for '0' to name an object that is not the same object as itself.) So there can be no such interpretation. That is, PA is UNSATISFIABLE: it is logically impossible for an interpretation to make all the axioms of PA true; it is logically impossible for PA to have a model. So

$$PA \vDash 0 \neq 0$$

and, hence, by the Completeness Theorem,

$$PA \vdash 0 \neq 0.$$

That is, PA is INCONSISTENT.[6] We reached this conclusion by first supposing that PA completely captures our concept of the natural numbers. So our grand conclusion is: PA completely captures our concept of the natural numbers only if PA is inconsistent.

If PA is consistent, it is INCOMPLETE. (It is a bit ironic that the incompleteness of PA follows from Gödel's *completeness* theorem.) What is at issue here is *expressive* incompleteness. There are mathematically important properties of the natural numbers that cannot even be expressed in the language of PA. Even if we use infinitely many sentences of PA, we cannot assert that every natural number is standard nor, to take another example, can we assert that each natural number is greater than only finitely many natural numbers.

Exercise 2.9 *Sketch a compactness argument showing that* PA *allows for infinitely large numbers. It might be helpful to start by using the vocabulary of* PA *to define the "less than" symbol '$<$'. Feel free to define any other vocabulary you find useful. Feel free, too, to assume that* PA *is consistent. You may, if you wish, fill in all the details of the argument, but I am only asking for the general idea.*

2.4 Incompleteness 2: Representability

In the intended interpretation of PA, the PA-NUMERAL for a natural number n consists of an occurrence of '0' preceded by n occurrences of 'S'. When we use the informal variable 'n' to make general claims about natural numbers, we can use the informal variable '$\mathbf{n}$' to make general claims about the corresponding PA-numerals. For example, we might say: if n is a natural number, then the PA-numeral $\mathbf{n}$ consists of an occurrence of '0' preceded by n occurrences of 'S'.

---

[6] Here are three equivalent definitions of inconsistency: an inconsistent theory is one that proves a logical absurdity such as '$0 \neq 0$'; an inconsistent theory is one that proves a sentence ϕ and its negation $-\phi$; an inconsistent theory is one that proves every sentence in its language.

Suppose f is a one-place, primitive recursive, characteristic function. By "primitive recursive", we mean essentially what we meant in the preceding chapter with any modifications necessary to accommodate the number zero. A one-place function takes natural numbers one at a time as inputs. A characteristic function has only two possible outputs: zero and one.

Suppose $\phi(x)$ is a PA-formula with free occurrences of 'x' and no free occurrences of any other variable. Then $\phi(x)$ is said to REPRESENT f in PA if it has the following properties. If n is a natural number and $f(n) = 1$, then PA proves the sentence that results when we replace every free occurrence of 'x' in $\phi(x)$ with an occurrence of the PA-numeral for n. That is:

$$f(n) = 1 \implies \text{PA} \vdash \phi(\mathbf{n}).$$

If n is a natural number and $f(n) = 0$, then PA proves the sentence that results when we replace every free occurrence of 'x' in $-\phi(x)$ with an occurrence of the PA-numeral for n. That is:

$$f(n) = 0 \implies \text{PA} \vdash -\phi(\mathbf{n}).$$

If f is represented in PA by a PA-formula, then we say that f is REPRESENTABLE in PA.

Each characteristic function answers a yes-or-no question about the natural numbers. It answers "yes" by returning the value 1. It answers "no" by returning the value 0. (In Chap. 1, || was "yes", while | was "no".) Suppose the characteristic function f is represented in PA by the formula $\phi(x)$. By feeding f a number n, we can learn whether PA thinks n has the property expressed by $\phi(x)$. f will answer "yes" only if PA proves $\phi(\mathbf{n})$. f will answer "no" only if PA refutes $\phi(\mathbf{n})$.

Here is an example. We can infer from Definition 1.14, that there is a primitive recursive characteristic function $\mathfrak{O}$ that behaves as follows.

$$\mathfrak{O}(n) = \begin{cases} 1 & \text{if } n \text{ is odd} \\ 0 & \text{otherwise.} \end{cases}$$

$\mathfrak{O}$ tells us whether the numbers we feed it are odd. If we want to find a PA-formula that represents f, we should look for one that somehow expresses the property of oddness. If $\mathfrak{O}$ were represented in PA by $\phi(x)$, then PA would prove each of the sentences

$$\phi(S0), \quad \phi(SSS0), \quad \phi(SSSSS0), \quad \phi(SSSSSSS0), \ldots$$

and would disprove each of the sentences

$$\phi(0), \quad \phi(SS0), \quad \phi(SSSS0), \quad \phi(SSSSSS0), \ldots$$

That is, PA would prove $\phi(\mathbf{n})$ whenever n is odd and would prove $-\phi(\mathbf{n})$ whenever n is even. Can we identify such a formula $\phi(x)$? Well, does the vocabulary of PA allow us to say that a number x is odd? Of course it does. We just say that x is not a multiple of two:

$$\forall y \; (y \cdot SS0) \neq x.$$

As it turn out, this formula really does represent f because PA proves each of the sentences

$$\forall y \; (y \cdot SS0) \neq S0, \; \forall y \; (y \cdot SS0) \neq SSS0, \; \forall y \; (y \cdot SS0) \neq SSSSS0, \; \ldots$$

and disproves each of the sentences

$$\forall y \; (y \cdot SS0) \neq 0, \; \forall y \; (y \cdot SS0) \neq SS0, \; \forall y \; (y \cdot SS0) \neq SSSS0, \; \ldots$$

So, when we offer f the number n and f responds "yes", we do not just learn that n is odd: we learn that PA thinks n is odd. When we offer f the number n and f responds "no", we do not just learn that n is even: we learn that PA thinks n is even.

Here is another example. We know from Definition 1.15, that there is a primitive recursive characteristic function $\mathfrak{P}$ that behaves as follows.

$$\mathfrak{P}(n) = \begin{cases} 1 & \text{if } n \text{ is prime} \\ 0 & \text{otherwise.} \end{cases}$$

Exercise 2.10 *Identify a* PA-*formula that represents* $\mathfrak{P}$ *in* PA.

It is no fluke that $\mathfrak{O}$ and $\mathfrak{P}$ are representable in PA. Gödel proved in 1930 that *all* of our primitive recursive functions f are representable in PA.[7] So if

$$f(\mathfrak{a}) = 1$$

is a true $\varPi_1^0$ sentence, there is a formula $\phi(x)$ that represents f in PA and, so, PA proves each of the sentences

$$\phi(0), \; \phi(S0), \; \phi(SS0), \; \phi(SSS0), \; \ldots$$

Since PA does not guarantee that

$$0, \; S0, \; SS0, \; SSS0, \; \ldots$$

are the only objects in the range of its bound variables, we should hesitate to infer that PA will prove

---

[7] See Gödel [6]; English translation in Gödel [7, pp. 145–195], and van Heijenoort [10, pp. 596–616].

$$\forall x \; \phi(x).$$

Indeed, Gödel identified primitive recursive functions f represented by PA-formulas $\phi(x)$ such that, if PA is consistent, it is true that $f(\mathfrak{a}) = 1$, but $\forall x \; \phi(x)$ is not provable in PA. That is, Gödel actually identified formulas $\phi(x)$ with the following odd property. PA proves each of the sentences

$$\phi(0), \; \phi(S0), \; \phi(SS0), \; \phi(SSS0), \; \ldots$$

and, so, confirms that each *standard* number satisfies $\phi(x)$. Yet, if PA is consistent, PA is unable to confirm that every number, every object in the range of its bound variables, satisfies $\phi(x)$ (If PA is *inconsistent* it "confirms" everything: every PA-sentence is a PA-theorem).

Suppose, on the other hand, that $f(\mathfrak{a}) = 1$ and PA $\vdash -\forall x \; \phi(x)$. Then, by Theorem 2.2, PA $\vDash -\forall x \; \phi(x)$ and, hence, every model of PA makes $-\forall x \; \phi(x)$ true (since no interpretation makes all the PA-axioms true and $-\forall x \; \phi(x)$ false). So every model of PA makes $\forall x \; \phi(x)$ false. However, since $\phi(x)$ represents f in PA, Theorem 2.2 implies that every model of PA makes each of the sentences

$$\phi(0), \; \phi(S0), \; \phi(SS0), \; \phi(SSS0), \; \ldots$$

true. This would mean that every model of PA includes a non-standard number in the range of PA's bound variables (a number that makes $\forall x \; \phi(x)$ false). That is, PA would have no STANDARD MODELS and, hence, would be logically incompatible with our concept of the natural numbers. Our grand conclusion: if PA is compatible with our concept of the natural numbers, then PA does not *disprove* $\forall x \; \phi(x)$.

Suppose, now, it is logically possible for PA to have a standard model (a model in which every object in the range of PA's bound variables is named by a PA-numeral). Then Gödel has shown us how to identify primitive recursive functions f represented by PA-formulas $\phi(x)$ such that (1) it is true that $f(\mathfrak{a}) = 1$, but (2) $\forall x \; \phi(x)$ is neither provable nor refutable in PA. PA-sentences that are neither provable nor refutable in PA are said to be UNDECIDABLE in PA. We shall now consider what sort of an f would allow us to establish undecidability.

We first note that we can use natural numbers to *code* formulas of PA. Let's code some of the symbols of PA as follows.

$-$	$\to$	$\forall$	$=$	x	y	S	0	$+$	$\cdot$	$($	$)$
1	2	3	4	5	6	7	8	9	10	11	12

Now consider the PA-sentence '$-0 = 0$'. To code this sentence, first replace each symbol with its code:

$$- 0 = 0$$
$$\downarrow \ \downarrow \ \downarrow \ \downarrow$$
$$1 \ 8 \ 4 \ 8$$

Now take the product of the first four prime numbers raised to the powers 1, 8, 4, 8:

$$2^1 \times 3^8 \times 5^4 \times 7^8.$$

This number

$$47,278,574,201,250$$

is the GÖDEL NUMBER of the PA-sentence '$-0 = 0$'.[8] Working the other direction, if you were presented with the number 47,278,574,201,250, you could ask your calculator or computer to factor it. Then, noting the exponents and consulting our table of symbol codes, you could reconstruct the PA-sentence '$-0 = 0$'.

Exercise 2.11 *Decode the Gödel number* $2,349,101,964,825,000$.

Exercise 2.12 *Decode the Gödel number*

$$96,860,719,328,790,117,762,174,283,536,741,369,039,814,728,960,000.$$

We can use Gödel numbering to code sequences of PA-formulas. For example, if ϕ and ψ are PA-formulas with codes #ϕ and #ψ, we can let

$$2^{\#\phi} \times 3^{\#\psi}$$

code the ordered pair $< \phi, \psi >$. (We already used this trick in Chap. 1.) We can use similar techniques to code formalized PA-proofs. Suppose we have done so. Gödel figured out how to identify a primitive recursive function g, represented in PA by a PA-formula $\gamma(x)$, that behaves as follows.[9]

---

[8] The corresponding PA-numeral consists of a single occurrence of '0' preceded by 47,278,574,201,250 occurrences of 'S'. If you were to produce a token of this numeral, it would be about 100 million km long. That is about two thirds the distance from the Earth to the Sun or about 2,500 times the circumference of the Earth. This suggests that it may be naive to think of PA-numerals as actual physical objects.

[9] For a readable discussion of Gödel's construction, see Nagel and Newman [8]. Another helpful resource on this and other issues of interest to us is George and Velleman [5]. It might help you wrap your brain around Gödel's proof if you read $\gamma(n)$ as "n does not code a PA-proof of **G**" where **G** is a certain extra-special sentence of PA. Then $\forall x \ \gamma(x)$, the universal generalization of $\gamma(n)$, says that no natural number codes a PA-proof of **G**. Now it so happens that $\forall x \ \gamma(x)$ *is* **G**. So **G** says of itself that it is not provable in PA. A PA-proof of **G** would prove that **G** is not provable in PA: a strange situation, to say the least. Of course, all this is a bit sloppy. $\gamma(n)$ does not say *anything* unless we interpret it. Furthermore, under the intended interpretation it does not say anything about PA-proofs: it only refers to natural numbers. But the road to clarity is sometimes paved with slop.

$$\mathfrak{g}(n) = \begin{cases} 1 & \text{if } n \text{ does not code a proof in PA of } \forall x \ \gamma(x) \\ 0 & \text{otherwise.} \end{cases}$$

Since $\gamma(x)$ represents $\mathfrak{g}$ in PA we also know:

$$\mathfrak{g}(n) = 1 \implies \mathsf{PA} \vdash \gamma(\mathbf{n})$$

$$\mathfrak{g}(n) = 0 \implies \mathsf{PA} \vdash -\gamma(\mathbf{n}).$$

If we offer $\mathfrak{g}$ the number n and $\mathfrak{g}$ responds "yes", then n does *not* code a PA-proof of the generalization $\forall x \ \gamma(x)$, but PA does prove the instance $\gamma(\mathbf{n})$. The other case is more interesting: if we offer $\mathfrak{g}$ the number n and $\mathfrak{g}$ responds "no", then n codes a PA-proof of $\forall x \ \gamma(x)$ and PA proves $-\gamma(\mathbf{n})$. But it is not coherent to say that every number has a certain property and, also, that a particular number does not. So if $\mathfrak{g}$ were ever to say "no", that would mean PA is inconsistent. Let us run through this argument more carefully. Suppose $\mathsf{PA} \vdash \forall x \ \gamma(x)$. Then we can pick a natural number k that codes a PA-proof of $\forall x \ \gamma(x)$. Note that $\mathfrak{g}(k) = 0$. So, since $\gamma(x)$ represents $\mathfrak{g}$, $\mathsf{PA} \vdash -\gamma(\mathbf{k})$. But, since $\forall x \ \gamma(x)$ is derivable in PA, $\mathsf{PA} \vdash \gamma(\mathbf{k})$. That is, PA is inconsistent. Our conclusion: if PA is consistent, then PA does not prove $\forall x \ \gamma(x)$. Suppose, now, it is logically possible for PA to have a standard model. Then, by Exercise 2.6, PA is consistent. So no natural number codes a PA-proof of $\forall x \ \gamma(x)$ and, hence, the Π_1^0-sentence

$$\mathfrak{g}(\mathfrak{a}) = 1$$

is true. So PA proves each of the sentences

$$\gamma(0), \ \gamma(S0), \ \gamma(SS0), \ \gamma(SSS0), \ \ldots$$

and, hence, by our earlier reasoning, PA does not disprove $\forall x \ \gamma(x)$ (since, otherwise, every model of PA would have to feature a non-standard number witnessing to the falsehood of $\forall x \ \gamma(x)$). Our grand conclusion: if it logically possible for PA to have a standard model, then $\forall x \ \gamma(x)$ is undecidable (neither provable nor refutable) in PA.

We earlier saw that PA suffers from a kind of *expressive* incompleteness: there are mathematically important properties of the natural numbers that cannot be expressed in the language of PA. We now see that PA suffers from a kind of *proof-theoretic* incompleteness: there are questions expressible in the language of PA that cannot be settled by a proof or refutation in PA. That is, there are cases where PA's expressive resources are up to snuff, but its capacity to supply proofs is not.

Here is another example of an undecidable sentence. There is a primitive recursive function $\mathfrak{s}$, represented in PA by a PA-formula $\sigma(x)$, that behaves as follows.

$$\mathfrak{s}(n) = \begin{cases} 1 & \text{if } n \text{ does not code a proof in PA that } 0 \neq 0 \\ 0 & \text{otherwise.} \end{cases}$$

Suppose PA is consistent. That is, suppose PA does not prove the absurdity '0 ≠ 0'. Then no natural number codes a PA-proof of '0 ≠ 0' and, hence, the Π_1^0-sentence

$$\mathfrak{s}(\mathfrak{a}) = 1$$

is true. So PA proves each of the sentences

$$\sigma(0), \ \sigma(S0), \ \sigma(SS0), \ \sigma(SSS0), \ \ldots$$

and, hence, if it is logically possible for PA to have a standard model, PA does not disprove $\forall x \ \sigma(x)$. It turns out that PA does prove

$$(\forall x \ \sigma(x) \to \forall x \ \gamma(x)).$$

So, if PA proved $\forall x \ \sigma(x)$, it would also prove $\forall x \ \gamma(x)$. We conclude: if it is logically possible for PA to have a standard model, then $\forall x \ \sigma(x)$ is undecidable in PA.

Suppose there is a PA-formula $\chi(x)$ that somehow expresses the idea that x codes a PA-proof of '0 ≠ 0'. If $\chi(x)$ is really doing a good job of expressing that idea inside PA, we would expect that, for each natural number n, if n does not code a PA-proof of '0 ≠ 0', then PA $\vdash -\chi(\mathbf{n})$ (that is, PA will refute the sentence asserting that n *does* do the coding). Suppose this is the case.

Exercise 2.13 *Prove that:* PA $\vdash \forall x \ - \chi(x)$ *only if* PA *is inconsistent.* (*You might start by considering the relationship between the formula* $-\chi(x)$ *and the function* $\mathfrak{s}$.)

Exercise 2.14 *Suppose that, for each natural number n,* PA $\vdash -\chi(\mathbf{n})$ *only if n does not code a* PA-*proof of* '0 ≠ 0'. *Show that this implies the consistency of* PA.

Exercise 2.13 shows: if PA allowed us to *say* that no natural number codes a PA-proof of '0 ≠ 0', PA would not allow us to *prove* this unless it allowed us to prove *everything*. So we will not be able to use a PA-proof to demonstrate the consistency of PA. More generally, we will not be able to prove the consistency of PA using methods formalizable in PA.

Note that a PA-proof of PA's consistency would not be as pointless as it might first appear. The PA-proof would use only finitely many axioms of PA: we would be relying on only finitely many axioms to show that no combination of the infinitely many PA-axioms proves an absurdity. Those finitely many axioms might have had some property that made their consistency evident or, at least, more evident than the consistency of PA as a whole. So the PA-proof might have provided a non-circular reason for believing PA consistent. Alas, we now recognize that this is not to be.

2.5 Why Fret About Consistency?

Hilbert thought that mathematicians frequently devote time and talent to proofs of sentences that have no content: sentences that may seem to state something, but really state nothing. He thought it important to *justify* this practice in some way. He recognized that *consistency proofs* could provide an especially powerful justification. Here is why.

Suppose f is a primitive recursive characteristic function represented in PA by $\phi(x)$. If $f(n) = 0$, then PA $\vdash -\phi(\mathbf{n})$. So if PA is consistent and PA $\vdash \phi(\mathbf{n})$, then $f(n) = 1$ (since, otherwise, PA would prove both $\phi(\mathbf{n})$ and $-\phi(\mathbf{n})$). So if PA is consistent and PA $\vdash \forall x \; \phi(x)$, then

$$f(0) = 1, \quad f(1) = 1, \quad f(2) = 1, \quad \ldots$$

are all true and, indeed, the Π_1^0 sentence

$$f(\mathfrak{a}) = 1$$

is true. If we are convinced that PA is consistent, we can feel free to use the machinery of PA to verify Π_1^0 sentences. We could think of PA as a trustworthy oracle. If the great oracle PA says that $\forall x \; \phi(x)$ is true, then that settles it: it really is true that $f(\mathfrak{a}) = 1$. The same argument applies to any theory in which all the PA-axioms are derivable: if the theory is consistent, we can use it to verify Π_1^0 statements. Let your imagination run wild: as long as your fanciful tales are not fundamentally incoherent, we can use them to verify propositions such as Fermat's Last Theorem and Goldbach's Conjecture. What matters is consistency, not truth.

If we have good reason to believe PA consistent, then we have good reason to use PA to prove theorems. If a theorem itself corresponds to a Π_1^0 sentence, we will verify that sentence. Otherwise, we will obtain results that (meaningless or not) may contribute to other proofs and, so, help us verify Π_1^0 sentences. Note that we need a *good reason* for believing PA consistent. A *mathematical proof* would certainly be a good reason. I note, though, that not every good reason in mathematics is supplied by a proof. (We might have very good reasons for adopting axioms, though axioms are statements we accept without proof.) GERHARD GENTZEN (1909–1945) *did* prove the consistency of PA using techniques not formalizable in PA.[10] If you are already convinced that each axiom of PA is a true statement about the natural numbers, you may feel no need for such a proof. After all, a collection of *true* sentences cannot be inconsistent. Indeed, sentences that *could* all be true cannot be inconsistent. I suppose a reasonable person could, nonetheless, question the consistency of PA. However, the logician SOLOMON FEFERMAN reports that, among today's mathematicians, "the number who doubt that PA is consistent is vanishingly small".[11]

---

[10] See Gentzen [3]; English translation in Gentzen [4, pp. 132–213].

[11] See Feferman [1, p. 192].

As I already mentioned, the above reasoning applies to any theory that extends PA, any theory in which all the PA-axioms are derivable. If we have good reason to believe such a theory consistent, we have good reason to use the theory to prove theorems. In the next chapter, we begin our study of an important family of such theories: set theories.

2.6 Solutions of Odd-Numbered Exercises

2.1 No. Yes. Yes. No. No. Yes.

2.3 A1 says that, no matter what y is, $(y + 0) = y$. So, in particular, $(0 + 0) = 0$. Suppose, as an inductive hypothesis, that $(0 + x) = x$. Then $S(0 + x) = Sx$. A2 says that, no matter what x and y are, $(y + Sx) = S(y + x)$. So, in particular, $(0 + Sx) = S(0 + x)$ and, hence, $(0 + Sx) = Sx$. We have shown

$$((0 + x) = x \rightarrow (0 + Sx) = Sx).$$

Since we have not used any special information about x, x could be *anything*. That is, $\forall x((0 + x) = x \rightarrow (0 + Sx) = Sx)$. One of our induction axioms assures us that

$$((0 + 0) = 0 \rightarrow (\forall x((0 + x) = x \rightarrow (0 + Sx) = Sx) \rightarrow \forall x (0 + x) = x)).$$

So $\forall x(0 + x) = x$.

2.5 Suppose $\Gamma \vDash \psi$. Then, by the completeness theorem, $\Gamma \vdash \psi$. That is, there is a formal proof whose conclusion is ψ and whose premises are members of Γ. Let the members of Γ' be the finitely many members of Γ that appear in the proof. Then $\Gamma' \vdash \psi$ and, hence, by the soundness theorem, $\Gamma' \vDash \psi$.

2.7 Suppose PA $\nvdash 0 \neq 0$. Then, by the completeness theorem, PA $\nvDash 0 \neq 0$. That is, it is logically possible for an interpretation to make all the axioms of PA true and '$0 \neq 0$' false.

2.9 We introduce the "less than" symbol by stipulating that $\ulcorner \alpha < \beta \urcorner$ is an abbreviation of $\ulcorner -\forall z \, (\alpha + Sz) \neq \beta \urcorner$. (The idea is that $x < y$ if and only if you can get to y by adding some non-zero number to x.) Let the members of C be the sentences

$$0 < c, S0 < c, SS0 < c, SSS0 < c, \ldots$$

Let C^* be a finite subset of C. Each member of C^* is an inequality $\ulcorner \alpha < c \urcorner$ where α is a PA-numeral. Since C^* is finite we can pick a PA-numeral β longer than all those α's. We are assuming that PA is consistent. So, by the completeness theorem and Exercise 2.7, it is logically possible for PA to have a model. Suppose we have picked such a model. That model will assign an object to β. Extend the model by assigning the same object to 'c'. Our new model will make $\ulcorner c = \beta \urcorner$ true. If our new model makes an inequality $\ulcorner \alpha < c \urcorner$ in C^* false, then (since our model thinks 'c' and β name the same thing) it will make the inequality $\ulcorner \alpha < \beta \urcorner$ false where α and β are PA-numerals, the latter longer than the former. We can show that this is impossible. (You might enjoy working out the details.) So, in fact, our new model makes all the members of C^* true. More generally, if C^* is any finite subset of C, it is logically possible for PA $\cup$ C^* to have a model. If every member of C were true, then c would exceed every finite value and, so, would deserve to be called infinitely large. Suppose PA rules out infinitely large numbers. Then the combination PA $\cup$ C is incoherent and, so, can have no models. Since no models are possible,

$$\text{PA} \cup C \vDash 0 \neq 0.$$

Use the compactness theorem to pick a finite subset of C such that

$$\text{PA} \cup C^* \vDash 0 \neq 0.$$

Then PA$\cup C^*$ cannot have a model—contrary to our earlier result. So we were wrong to suppose that PA rules out infinitely large numbers.

2.11 $\forall x \; x = x$.

2.13 First note the following

$$\mathfrak{s}(n) = 1 \implies n \text{ does not code a proof in PA that } 0 \neq 0$$
$$\implies \text{PA} \vdash -\chi(\mathbf{n})$$
$$\mathfrak{s}(n) = 0 \implies n \text{ codes a proof in PA that } 0 \neq 0$$
$$\implies \text{PA proves every sentence of PA}$$
$$\implies \text{PA} \vdash - - \chi(\mathbf{n}).$$

So $-\chi(x)$ represents $\mathfrak{s}$ in PA and, hence, by our earlier reasoning, if PA proved $\forall x - \chi(x)$, it would also prove $\forall x \; \gamma(x)$. But PA does *that* only if it is inconsistent.

References

1. Feferman, S. (1998). *In the light of logic*. New York: Oxford University Press.
2. Frege, G. (1879). *Begriffsschrift*. Halle: Louis Nebert.
3. Gentzen, G. (1936). Die Widerspruchsfreiheit der reinen Zahlentheorie. *Mathematische Annalen, 112*, 493–565.
4. Gentzen, G. (1969). *The collected papers of Gerhard Gentzen*. Amsterdam: North-Holland.
5. George, A., & Velleman, D. J. (2002). *Philosophies of mathematics*. Oxford: Blackwell.

6. Gödel, K. (1931). Über formal unentscheidbare Sätze der Principia Mathematica und verwandter Systeme I. *Monatshefte für Mathematik und Physik, 38,* 173–98.
7. Gödel, K. (1986). *Collected works* (Vol. 1). New York: Oxford University Press.
8. Nagel, E., & Newman, J. R. (1958). *Gödel's proof.* New York: New York University Press.
9. Quine, W. V. O. (1981). *Mathematical logic.* Cambridge, MA: Harvard University Press.
10. van Heijenoort, J. (Ed.). (1967). *From Frege to Gödel.* Cambridge, MA: Harvard University Press.
11. Wang, H. (1957). The axiomatization of arithmetic. *Journal of Symbolic Logic, 22,* 145–58.

Chapter 3
Hereditarily Finite Lists

3.1 What Sets Are Not

Professors really do mean well. If they are excited by the mathematical theory of sets, they will want others to be excited too. They will try to entice their students, to make their initial experience as pleasant as possible. They may say, "Sets are like . . .," with the blank occupied by some reference to a non-threatening item of everyday experience. "Sets are like boxes." "A set is like a flock of geese." "A set is what you get when you bunch stuff together."

Professors do mean well. When they say, "Sets are like . . .," their intentions are pure and noble. They want to communicate something interesting and important. By putting students at ease, they hope to communicate it much more effectively. In the face of such good intentions, I should probably keep my mouth shut. But here it is: the students are being hoodwinked. I am confident there are no objects of everyday experience that closely resemble mathematical sets. The popular incantations, "Sets are like . . .," may be psychologically and pedagogically beneficial, but they all fall apart under critical scrutiny. When treated as factual assertions, rather than poetry, they turn out to be as flimsy as wet paper.[1] Here is one example.

Professor: Sets are like boxes. The things in a box are like the members of a set.

Student: Here are two boxes. Everything in one is in the other, because, in fact, they are both empty. If sets are like boxes, then I guess different sets can have the same members.

Professor: Well, no. A basic principle of set theory is that sets with the same members are the same.

Student: So if I want to arrange for some boxes to behave like sets, I have to make sure that different ones have different contents.

Professor: Yes.

---

[1] For references and further discussion, see Pollard [11], Chap. 3.

S. Pollard, *A Mathematical Prelude to the Philosophy of Mathematics*,
DOI: 10.1007/978-3-319-05816-0_3,
© Springer International Publishing Switzerland 2014

Student: What if I put my pen in this box? It's the same box, but now it has something in it that wasn't there before. If sets are like boxes, then I guess sets with different members can be the same.

Professor: Certainly not. It's a basic principle of mathematical logic that sets with different members are different.

Student: So my boxes will behave like sets as long as I seal them all up and don't let anyone shift things from one to another.

Professor: Yes.

Student: I put my pen in box *A*. Now I'm putting box *A* in box *B*. So now my pen is also in box *B*. If sets are like boxes, then I guess anything that's a member of a member of a set *z* will itself be a member of *z*.

Professor: Well, no. If *x* is a member of *y* and *y* is a member of *z*, it doesn't follow that *x* is a member of *z*.

Student: So the things in a box will be like the members of a set as long as we take "in" to mean something like "immediately inside of." My pen is in box *B*, but is not immediately inside of it because there is another box between the pen and box *B*.

Professor: That sounds right.

Student: OK, I've sealed up all my boxes. I've made sure that different ones have different contents. I'm taking "in" to mean "immediately inside of." But now I notice something: my pen cannot be immediately inside of more than one box. If sets are like boxes, then I guess nothing belongs to more than one set.

Professor: No, no! We can prove that every set belongs to infinitely many sets.

And so on.

I am going to try my best not to hoodwink you. I am *not* going to say that mathematical sets are closely similar to any items of everyday experience. I know of no such items. I *will* insist that sets are, in several important respects, like certain things we can readily understand on the basis of everyday experience. The point is that experience with everyday objects can help us understand things that are not objects of everyday experience. We are going to discuss some things, known as "unranked list-types," that are not physical objects.[2] You will not bump against them as you move through space. You cannot find one in your pocket or under a rock or even in a distant galaxy. Nonetheless, your experience of some of the things you do bump against (such as paper lists) will help you understand unranked list-types. Even if you end up a skeptic about non-physical beings, certain experiences will help you understand what unranked list-types *would* be like if they *were* to exist. This, in turn, will help you understand mathematical sets. So, yes, everyday experience can help you understand mathematical sets. I do not deny *that*, but I do insist that the path from physical realities to mathematical idealities is longer than some professors admit.

To repeat: mathematical sets are, in several important respects, like certain things, unranked list-types, that we can readily understand on the basis of everyday

---

[2] You already encountered the type/token distinction in Chap. 1. The relation between a single list-type and its many paper or pixel tokens is like the relation between a single numeral-type and the many numeral-tokens of that type.

experience. There are also important dissimilarities and I will be careful to point out various ways that unranked list-types are unlike sets. Our starting point is the commonplace remark that you can identify a finite set by listing its members. If the list displays all the essential properties of the set, why not ignore the set and concentrate on the list?[3]

3.2 List-Types

Suppose I decide to list the three Beethoven piano sonatas I most admire. You decide to make a list of your own. I show you this:

opus 14, #2

opus 53

opus 109

You show me this:

opus 109

opus 53

opus 14, #2

As long as it is understood that our lists are unranked, it would be correct for one of us to say, "Hey, we have the same list." Now the physical objects we have displayed, the paper lists, are clearly distinct. Mine is here; yours is over there. Yet we say we have the same list. So the one-and-the-same list we share is not a paper list. The philosophers would say it is a LIST- TYPE. This terminology should not be too hard to swallow since we ourselves might say our two paper lists are "of the same type." That type or species itself is the list we share. From now on let us use the word 'list' to refer to unranked list-types. Remember: the one list we share, the list-type, is not

---

[3] This is a strategy mathematicians sometimes employ. For example, see Edwards [4], p. 139, for a treatment of multi-sets as lists. A multi-set is a set whose members can occur more than once. For a theory of *ordered* lists whose entries can occur more than once, see Deiser [3]. Such lists are not unusual. For instance, here is how the list of Wimbledon Gentlemen's Singles champions begins.

1. Spencer Gore
2. Frank Hadow
3. John Hartley
4. John Hartley

This is not a *ranking*. (A gentleman cannot outrank himself.) In mathematical parlance, it is a TUPLE (in particular, a 4-tuple). The lists we will be discussing in this chapter are not just unranked; they are not even ordered: they are not tuples. Among ranked lists there are some with ties and some without. Our lists are not of either sort.

to be identified with the marked-up piece of paper in your hand. Go ahead; tear up your paper if you want. You will not be tearing up our one shared list.

This intangible, invisible list of ours may seem a bit mysterious. If we think about it too hard, we are likely to churn up all sorts of puzzles and get all sorts of mental cramps. None of this need get in the way of our current project. We do not even need to believe that such lists exist. I am not trying to establish the existence of anything. I am just trying to convey what it would be like if certain structures were to exist. All this requires is a basic understanding of some fairly ordinary talk about lists and a notion of what it would be like for that talk to be true when given a rather literal-minded interpretation. If the outcome is a grasp of the right structures, it will not matter if our understanding of lists is imperfect and gives us mental cramps when we probe it in certain ways.

We are taking seriously the idea that your list is the same as mine. How did we figure out that our lists are the same? Well, we noticed that we listed the same things. There is a general principle at work here: lists (that is, unranked list-types) are the same if they list the same items.

What are these items? We talk about things *appearing* on lists. This should not be confused with certain marks appearing (that is, being physically present) on a piece of paper. One could note, quite correctly, that the Waldstein sonata appears on my list. That is because the Waldstein sonata is Beethoven's opus 53 and opus 53 appears on my list. The words 'Waldstein sonata' do not appear on my list; no marks forming these words appear on my list; but the Waldstein sonata does appear. I have not listed phrases that refer to Beethoven sonatas (though I did write down such phrases). I have listed Beethoven sonatas. The "listing" relation holds between a list and the things listed. One could list phrases that refer to Beethoven sonatas.

'Waldstein sonata'

'Beethoven's opus' 53

'Beethoven's 21st piano sonata'

The point is simply that this is different from listing piano sonatas. Note that the list

opus 14, #2

Waldstein sonata

opus 109

and the list

opus 14, #2

opus 53

opus 109

are the same because they list the same things. Here I am not announcing some profound discovery about the inner workings of the universe. I am just letting you know how I understand the expression 'unranked list-type'.

It is useful to have several ways of referring to the relationship between a list and an item listed. We say that y APPEARS ON x or, more briefly, x LISTS y. If x lists y, we also say that y is one of x's ENTRIES. Though it is not standard English, I say that a list's entries FORM that list. My three favorite Beethoven piano sonatas form a list: the very list we have been discussing. Some useful notation: we let $\{a, b, c, \ldots\}$ be the list whose entries are $a, b, c, \ldots$. So $a, b, c, \ldots$ form $\{a, b, c, \ldots\}$.

Lists can list lists. For example, I could list the first three lists discussed in this chapter.

My list of three piano sonatas

My list of three phrases

My list of three lists

This list has the odd property of listing itself. My first two lists do not list themselves. Could we list all the lists that do not list themselves? The mathematician ERNST ZERMELO (1871–1953) figured out that this is logically impossible.[4]

Exercise 3.1 *Show that it would be absurd for there to be a list that lists exactly those lists that do not list themselves.*

Zermelo's discovery should be a warning to us: assertions about the existence of lists can far too easily land us in a contradiction. We should handle such assertions with care.

Though it seems a bit of a stretch, we can imagine a list that lists nothing. I might undertake to list all the mass murderers I admire. When someone asks me how my list is coming along, I display a blank piece of paper. We could say there is no list of mass murderers I admire, since I admire no mass murderers. But it seems harmless to say there is such a list: a list with the odd property that it has no entries. ('There is nothing on your list' is a perfectly acceptable sentence of English.) Since lists with no entries all have the same entries, they are all the same list: the one-and-only BLANK LIST. By displaying my blank piece of paper, I am identifying my list as the blank list. More notation: we let $\emptyset$ be the blank list.

---

[4] Kanamori [7], Landini [9], Rang and Thomas [12] give some of the history.

3.3 Precursors

People can do things together that none of them can do alone. Fifteen FBI agents can surround a house. It is not even clear what it would *mean* for one of them to do so.[5] Lists behave the same way: they can have a property collectively that each of them lacks individually. For example, the first three lists I discussed in this chapter cooperate to form the third list I discussed. (That is, the first three lists are exactly the entries of the third list.) The first list does not, by itself, form the third list. Neither does the second nor even the third. (The third list is one of its own entries, but it is not all three of them. So it does not form itself.) The three lists have to work together to do this "forming." In this respect, they are like three musicians who can play a trio together, though none of them can do it solo.

Here is another example of a collective property. Consider the following lists: ∅, {∅}, {{∅}} (the blank list, the list whose only entry is the blank list, and the list whose only entry is the list whose only entry is the blank list). These lists are, collectively, ENTRY- CLOSED. That is, each of their entries is itself one of these very lists. The third list has one entry: {∅}. This is the second of the three lists. You can encounter it without going beyond the three. The second list has one entry: ∅. This is the first of the three lists. Again, you do not have to venture beyond the three to encounter this entry. The first list has no entries. (So it is easy to confirm that each of its entries is one of the three lists.) We can always fabricate some entry-closed lists by starting with a list, identifying each of its entries, identifying each entry of those entries, and so on until we run out of entries. For example, start with {{∅}}. This list has one entry: {∅}. This list, in turn, has one entry: ∅. At this point, we have run out of entries. So our procedure yields three lists ∅, {∅}, {{∅}} that, as we saw, are entry-closed. Note that the two lists ∅, {{∅}} are not, taken together, entry-closed, since {∅} is not one of them. The list ∅, just by itself, is entry-closed.

Exercise 3.2 *The one list* {∅, {{{∅}}, {∅}}} *is not, by itself, entry-closed. Starting with this list, identify lists until your lists, taken together, are entry-closed.*

Now consider any lists that are, collectively, entry-closed. Suppose y is one of them. Then each of y's entries is one of them. And each entry of an entry of y is one of them. And each entry of an entry of an entry of y is one of them. And so on. (More in a second about the phrase 'and so on'.) Let us refer to these entries of entries of entries ... of entries of y as PRECURSORS of y. Precursors have the following properties.

Proposition 3.1 *If some lists are entry-closed and y is one of them, then each precursor of y is one of them.*

Proposition 3.2 *The precursors of any list are, collectively, entry-closed.*

---

[5] In the folk song "Bad Man's Blunder," the bad man "got surrounded by a sheriff down in Mexico." I understand this to be a slightly odd way of saying that the sheriff arranged for the bad man to be surrounded, not that the sheriff did the surrounding all by himself.

If Proposition 3.2 is not evident, just note that any entry of one of y's precursors will itself be a precursor of y.

A list's precursors are its entries, the entries of its entries, the entries of the entries of its entries, and so on. This may seem to make sense. But is it really so clear what 'and so on' means here? I actually think it is clear. Others may disagree. Clear or not, it does not really matter: we can avoid 'and so on' altogether by using a trick invented by Gottlob Frege.[6] The idea is to turn Proposition 3.1 into our *definition* of 'precursor', in a way to be specified in a minute. A slight glitch: this definition will make each list a precursor of itself. Anyone who finds this offensive is encouraged to replace 'precursor' with a better term. (And, please, let me know what that term is.) A reason not to take offense: if we think of an entry of an entry of y as a second-order precursor of y and think of an entry of y as a first-order precursor of y, then a mathematician might find it natural to think of y itself as a zero-order precursor of y.

Now for Frege's definition. Suppose, no matter what entry-closed lists you might consider, x will appear among them if y does. We stipulate that such x's are the "Frege-precursors" of y. More formally:

Definition 3.1 x is a FREGE- PRECURSOR of y if and only if, given any entry-closed lists, if y is one of those lists, then so is x.

The Frege-precursors of y are what tireless immortal beings encounter when they start with y and persist in tracking down the entries of everything they encounter. We now hasten to show that we could live without this new term 'Frege-precursor': a list's Frege-precursors are exactly its precursors. We take this detour through Frege-precursors to confirm that we could introduce the term 'precursor' without using the phrase 'and so on'.

Proposition 3.1 implies that every precursor is a Frege-precursor. What about the converse? Could something be a Frege-precursor without being a precursor? Well, y is one of its own Frege-precursors. (Given any entry-closed lists, if y is one of them, then y is one of them.) But y will not necessarily be an entry of itself or an entry of an entry of itself or be embedded inside itself at all. So, as anticipated, the notion of Frege-precursor is a bit wider than our original notion of precursor. It does no harm, though, to expand the latter notion a tiny bit. A more important question: could something other than y be a Frege-precursor without being a precursor in our intended sense? That is not going to happen. For suppose x is a Frege-precursor of y. That is:

Given any entry-closed lists, if y is one of them, then so is x.

Proposition 3.2 assures us that y's precursors are, collectively, entry-closed. So the phrase 'given any' allows us to pick these: the precursors of y. That is:

If y is one of y's precursors, then so is x.

---

[6] Readers who like a challenge, might tackle Frege's own exposition in Frege [5]; English translation in van Heijenoort [13], pp. 5–82.

We have decided to let y be one of its own precursors. So x is a precursor of y and, more generally, something is a Frege-precursor of y if and only if it is a precursor of y. Apart from the slight glitch of making y one of its own precursors, Frege's definition captures the notion we hoped it would capture. If anyone wants to know what we meant earlier when we said "and so on," we can send them to Frege's definition. From now on, we will use Definition 3.1 as our definition of 'precursor' and will drop the 'Frege' part.

Exercise 3.3 *Say that your* ANCESTORS *are your parents, your parents' parents, your parents' parents' parents, and so on. Use Frege's technique to explain what an ancestor is without using the phrase 'and so on'.*

Exercise 3.4 *Use Definition 3.1 to show that $\emptyset$ is the only precursor of $\emptyset$.*

Exercise 3.5 *Use Definition 3.1 to show that the precursors of y's precursors are themselves precursors of y.*

Exercise 3.6 *Let "the Y-lists" be y itself and the precursors of y's entries. Show, first, that the Y-lists are entry-closed. Use this result and Definition 3.1 to show that each precursor of y is either y itself or a precursor of an entry of y.*

Exercise 3.7 *Use Definition 3.1 to show that if y lists the entries of its entries, then the precursors of y are y itself and y's entries.*

3.4 Logically Possible Lists

We have already seen that it is easy to produce a description of a list that turns out to be incoherent (as with the list of all lists that do not list themselves). Speaking a bit carelessly: there are lists that could not be. Our notion of precursor will help us explore what lists there could be. I suggest we start out with a particular list and see what lists it allows us to list. Take the list $\{\{\emptyset\}\}$. Since it has only three precursors we can easily list them: $\emptyset$, $\{\emptyset\}$, $\{\{\emptyset\}\}$. If a list had 10^{100} precursors, we human beings could not list them (at least, not in any sense of "list" I recognize). But this is because of our *physical* limitations. It does not seem *logically* impossible for the precursors of each list to be listed. Now when mathematicians have convinced themselves that something could be the case, they sometimes go ahead and assert that it is the case. This might be an expression of certainty. ("Not only could this be true; it definitely is true.") On the other hand, it might just be the first step in seeing what interesting things follow from the proposition asserted. In the latter spirit, we shall entertain a proposition about what lists there are.

Proposition 3.3 *The precursors of any list form a list.*

We hereby announce our intention to explore what the world would be like if such a thing were true.

Before we get carried away, though, notice that our principle seems too modest. If there is nothing fundamentally incoherent about listing *all* the precursors of a list, there should be nothing fundamentally incoherent about listing *some* of them. There are 2^3 lists that list only precursors of $\{\{\emptyset\}\}$. One of them lists none. Three list one; another three list two. One, as we have already seen, lists all three. This suggests a slightly stronger principle.

Proposition 3.4 *Any precursors of a list form a list.*

That is, if x is a list and $a, b, c, \ldots$ are all precursors of x, then $\{a, b, c, \ldots\}$ is a list.

Exercise 3.8 *List all the lists that list only precursors of* $\{\{\emptyset\}\}$.

You just listed all the lists of precursors of $\{\{\emptyset\}\}$. If you could list all of them, surely you could have listed some of them. Without bothering to identify all 2^8 lists of lists of precursors of $\{\{\emptyset\}\}$, let us go ahead and assert the corresponding principle.

Proposition 3.5 *Any lists of precursors of a list form a list.*

That is, if x is a list and $a, b, c, \ldots$ list only precursors of x, then $\{a, b, c, \ldots\}$ is a list. The idea behind this principle comes from the philosopher DAVID BENNETT.[7]

Should we now assert that any lists of lists of precursors of a list form a list and that any lists of lists of lists of precursors of a list form a list and so on forever? Not to worry: that will not be necessary. Our principles already imply all the propositions of this type we are going to need. Let us verify this. Here are the axioms we are going to use.

Axiom 3.1 Lists are the same if they have the same entries.

Axiom 3.2 There is a list with no entries.

Axiom 3.3 Any lists of precursors of a list form a list.

For the rest of this chapter, we will identify interesting things that follow from these axioms. Some new vocabulary will be useful.

Definition 3.2 A list is PURE if and only if all its precursors are lists.

The list $\{\emptyset, \{\text{Frege}\}\}$ is not pure because Frege is one of its precursors and (presumably) Frege is not a list. Your eight lists of precursors of $\{\{\emptyset\}\}$ are all pure. Precursors of a pure list are themselves lists of precursors of that pure list. (Each of them is a list and each of their entries is a precursor of the pure list.) So Axiom 3.3 yields the following theorem.

---

[7] See Bennett [2] (available via http://projecteuclid.org).

Theorem 3.1 *Any precursors of a pure list form a list.*

More narrowly:

Theorem 3.2 *Any entries of a pure list form a list.*

Definition 3.3 A PART of a list y is any list of all, some, or none of y's entries. (That is, x is a part of list y if and only if x is a list all of whose entries are entries of y.) A PROPER PART of y is any part of y other than y itself.

Note that each list is a part of itself (though not a proper part) and that the blank list is a part of every list. ($\emptyset$ is a part of list y because $\emptyset$ has no entries that are not entries of y.) Since each entry is a precursor, each part is a list of precursors. So Axiom 3.3 implies:

Theorem 3.3 *Any parts of a list form a list.*

In particular, *all* of a list's parts form a list.

Definition 3.4 Py is the list of y's parts.

Exercise 3.9 *Show that Py is pure if y is. (Remember Exercises 3.5 and 3.6.)*

It is a special case of Theorem 3.1 that all of a pure list's precursors, acting in concert, form a list.

Definition 3.5 If y is pure, then Sy is the list of y's precursors.

Exercise 3.10 *Show that Sy is pure if y is. (Remember Exercises 3.5 and 3.6.)*

Suppose y is pure. PSy is the list of all lists of y's precursors and $PPSy$ is the list of all lists of lists of y's precursors. Any entries of $PPSy$ form a list. So any lists of lists of precursors of a pure list form a list. We could repeat this argument as many times as we want. So, as long as we concentrate on pure lists, we'll have all the lists of lists of lists ... of lists of precursors we want.

Exercise 3.11 *Show that $Sy \neq \emptyset$.*

Exercise 3.12 *Show that Px is a proper part of Py only if x is a proper part of y.*

3.5 Numbers Can Be Lists

It seems likely that Axioms 3.1–3.3 are not logically absurd. It seems likely that they are consistent. The logician WILHELM ACKERMANN (1896–1962) figured out how to prove they are consistent.[8] Natural numbers can have a lot of internal structure that can be revealed in various ways: by taking the prime factorization, for example.

---

[8] See Ackermann [1]. For a more recent discussion, see Kaye and Wong [8] (available via http://projecteuclid.org).

This internal structure can be used to store information. Ackermann appreciated something that now may seem obvious: we can store a lot of information in natural numbers by exploiting their binary representation.

Suppose you are going to meet with Julia, Martin, and Yuri and need to know which of them are spies. I send you a very short message: "6." We have already agreed that you are to convert my message into binary, writing the digits under the three names taken in alphabetical order. Here is what you get.[9]

Julia	Martin	Yuri
1	1	0

We agreed that 1 means *yes* and 0 means *no*. So my message says that Julia and Martin are spies, but Yuri is not. That is, 6 stands for the list {Julia, Martin}.

Of course, numbers do not have to be lists of spies. They could be lists of lists. What list of lists is 6? To find out, first represent 6 in binary: 110. Since we have only three digits, 6 is going to list at most three lists. Our candidates are going to be the lists represented by the numbers 0, 1, and 2. (Yes, I realize we do not yet know what *these* lists are! Be patient. We will figure that out in a minute.) We do our decoding much as we did with the spies, though we now write down our candidate entries in descending numerical order.

List 2	List 1	List 0
1	1	0

Again, 1 means *yes* and 0 means *no*. So 6 says yes to 1 and 2, but no to 0.

List 2	List 1	List 0
1	1	0
Yes!	Yes!	No!

That is, $6 = \{1, 2\}$. What are 1 and 2? A number with n binary digits passes judgment on lists 0 through $n - 1$. So 1, with its one binary digit, passes judgment only on the first of our lists.

List 0
1
Yes!

1 says yes to 0. So $1 = \{0\}$. What is 0? That is easy: 0 is just one big *no*. It says no to the one list on which it passes judgment.

List 0
0
No!

---

[9] Recall that 6 is 110 in binary notation because $6 = 1 \cdot 2^2 + 1 \cdot 2^1 + 0 \cdot 2^0$.

Since 0 say no to 0, it lists nothing at all: $0 = \emptyset$. Since 1 says yes to 0 and to nothing else, 0 is the one entry of 1. That is, $1 = \{\emptyset\}$. What is 2? 2 is 10 in binary: yes to list 1 (that we have identified as $\{\emptyset\}$), no to list 0 (that we have identified as $\emptyset$).

$$\begin{array}{cc} \{\emptyset\} & \emptyset \\ 1 & 0 \\ \text{Yes!} & \text{No!} \end{array}$$

So $\{\emptyset\}$ is the one entry of 2. That is, $2 = \{\{\emptyset\}\}$. Now, finally, we know what 6 is. Since 6 is 110 in binary, it says yes to list 2 (that we have identified as $\{\{\emptyset\}\}$), yes to list 1 (that we have identified as $\{\emptyset\}$), and no to list 0 (that we have identified as $\emptyset$).

$$\begin{array}{ccc} \{\{\emptyset\}\} & \{\emptyset\} & \emptyset \\ 1 & 1 & 0 \\ \text{Yes!} & \text{Yes!} & \text{No!} \end{array}$$

So the two entries of 6 are $\{\{\emptyset\}\}$ and $\{\emptyset\}$. That is, $6 = \{\{\{\emptyset\}\}, \{\emptyset\}\}$.

Here is one more example. We shall do all our figuring in binary. What is 1010? Our decoding key is the following.

$$\begin{array}{cccc} \text{List 11} & \text{List 10} & \text{List 1} & \text{List 0} \\ 1 & 0 & 1 & 0 \end{array}$$

We already know that $10 = \{\{\emptyset\}\}$, $1 = \{\emptyset\}$, and $0 = \emptyset$. What is 11? 11 says yes to lists 1 and 0.

$$\begin{array}{cc} \{\emptyset\} & \emptyset \\ 1 & 1 \\ \text{Yes!} & \text{Yes!} \end{array}$$

So $11 = \{\{\emptyset\}, \emptyset\}$. This yields the following decoding scheme.

$$\begin{array}{cccc} \{\{\emptyset\}, \emptyset\} & \{\{\emptyset\}\} & \{\emptyset\} & \emptyset \\ 1 & 0 & 1 & 0 \\ \text{Yes!} & \text{No!} & \text{Yes!} & \text{No!} \end{array}$$

So $1010 = \{\{\{\emptyset\}, \emptyset\}, \{\emptyset\}\}$.

Exercise 3.13 *Decode the number* 18.

Ackermann's coding trick allows us to interpret claims about lists as claims about natural numbers. Axiom 3.1 says that numbers with the same binary representation are the same. Axiom 3.2 says there is a number with no 1's in its binary representation. Axiom 3.3 says something a bit obscure; but it too turns out to be true.

In the Ackermann interpretation, "lists" are just natural numbers. If such a list k is of the form $2^a + 2^b + 2^c + \cdots$ (where $a, b, c, \ldots$ are distinct natural numbers), then its "entries" are the exponents $a, b, c, \ldots$, with each of these representing a list. If

$$a = 2^{a_1} + 2^{a_2} + 2^{a_3} + \cdots$$

then $a_1, a_2, a_3, \ldots$ are all entries of a and, hence, are all entries of entries of k. A similar analysis of, say, a_1 could reveal entries of entries of entries of k. The rather bland idea behind the Ackermann reading of Axiom 3.3 is that this process cannot go on forever. For example, consider the "list" 9. We dissect this as follows.

$$9 = 2^3 + 2^0$$
$$\downarrow$$
$$3 = 2^1 + 2^0$$
$$\downarrow$$
$$1 = 2^0$$

The idea is to provide a binary representation of each non-zero exponent until we run out of such exponents. Since $k < 2^k$, the exponents get smaller at each step and, so, we have to run out of non-zero ones eventually. The "precursors" of 9 are 9 itself and the exponents that appear in this analysis. They are: 9, 3, 1, 0. If we want to list some of these precursors, there is nothing to stop us. For example, since $515 = 2^9 + 2^1 + 2^0$, it "lists" 9, 1, and 0. Similarly, there is nothing to stop us from listing any of these lists of precursors. So any lists of precursors of a list form a list.

Exercise 3.14 *Identify the precursors of* 12. *Identify a number that lists some lists of* 12's *precursors.* (*There are* 4, 294, 967, 296 *such numbers.*)

Axioms 3.1–3.3 *could* all be true because they *are* all true under the Ackermann interpretation. Logical absurdities are propositions that cannot be true. So the system consisting of Axioms 3.1–3.3 is not logically absurd. Or, to state the matter more carefully, our system is absurd only if arithmetic is absurd. The consistency of our system is at least as certain as the consistency of arithmetic. A reasonable person could consider this a proof of the consistency of our system.

3.6 Lists Can Be Numbers

We now have some idea of how our theory of lists could be developed inside number theory. We can also develop number theory inside our theory of lists. Recall that Sx is the list of x's precursors. We now introduce some new terminology.

Definition 3.6 Some lists are, collectively, S-CLOSED if and only if they satisfy the following condition: if a list x is one of them, then x's precursors form a list and that list too is one of them.

Definition 3.7 A list x is a NUMBER if and only if it satisfies the following condition: given any S-closed lists, if Ø is one of them, then so is x.

Numbers are what tireless immortal beings encounter when they start with $\emptyset$ and persist in applying the operation S to everything they encounter.

Exercise 3.15 *Prove that $\emptyset$ is a number.*

Our definition of number immediately yields an induction principle.

Theorem 3.4 *If some lists are S-closed and $\emptyset$ is one of them, then every number is one of them.*

In an ordinary inductive proof, we first show that 0 has a certain property. We then suppose that an arbitrary number n has the property and use this information to show that $n + 1$ also has the property. This establishes that every natural number has the property. Theorem 3.4 justifies a similar technique. First, show that $\emptyset$ has a certain property. Second, assume that an arbitrary number n has the property and use this information to show that Sn also has the property. Third, conclude that every number has the property. We use this form of induction to prove the next two theorems.

Theorem 3.5 *If n is a number, then Sn exists and is itself a number.*

Proof If x is pure, then, by Theorem 3.1, Sx exists and, by Exercise 3.10, it too is pure. So the pure lists are, collectively, S-closed. Furthermore, by Exercise 3.4, $\emptyset$ is pure. So, by Theorem 3.4, every number is pure (since $\emptyset$ is pure and purity is always transmitted from x to Sx). Suppose n is a number. Then, by Theorem 3.1, Sn exists. Since n is a number, you will find it among some S-closed lists whenever you find $\emptyset$ among them. So you will find Sn among some S-closed lists whenever you find $\emptyset$ among them. That is, Sn is a number. $\square$

Theorem 3.6 *Each number lists the entries of its entries.*

Proof $\emptyset$ lists all the entries of its entries for the same reason it lists all the even primes greater than 2: there are no such things. Suppose n lists the entries of its entries. Then, according to Exercise 3.7, the precursors of n are n itself and n's entries. Suppose Sn lists k and k lists j. Then k is either n itself or an entry of n. In either case, j is a precursor of n and, so, is listed by Sn. So Sn lists the entries of its entries. We have shown that $\emptyset$ satisfies the theorem and that Sn does whenever n does. Now we just apply Theorem 3.4. $\square$

Exercise 3.7 now yields the following.

Theorem 3.7 *If n is a number, then n and its entries form Sn (that is, the precursors of n are n itself and the entries of n).*[10]

Exercise 3.16 *Show that no number lists itself. (You might try induction.)*

Exercise 3.17 *Show that if m and n are numbers and $Sm = Sn$, then $m = n$.*

---

[10] In set theoretic notation: $Sn = n \cup \{n\}$.

Exercise 3.18 *Show that if m and n are numbers and n lists m, then Sn lists Sm. (You might try induction.)*

Exercise 3.19 *Show that every number other than ∅ lists ∅. (You might try induction.)*

∅ is playing the role of 0 here. $S∅$ is 1. $SS∅$ is 2. And so on. Each number is the list of prior numbers.

$$
\begin{aligned}
0 &= & ∅ \\
1 &= & \{0\} \\
2 &= & \{0, 1\} \\
3 &= & \{0, 1, 2\}
\end{aligned}
$$

That is, $Sn = \{0, \ldots, n\}$. The mathematician DIMITRY MIRIMANOFF (1861 – 1945) invented this beautiful technique for constructing objects that behave like the natural numbers.[11] On this basis, we can reproduce arithmetic using nothing but lists.

Since each number lists all smaller numbers and is listed by all larger ones, "number j lists number k" is equivalent to "k is smaller than j" and "j is larger than k." We have already shown that, as long as we are talking about numbers, the "listing" (or "larger than") relation is TRANSITIVE (Theorem 3.6) and IRREFLEXIVE (Exercise 3.16). We now show that it has two other important properties.

Theorem 3.8 WELL-FOUNDEDNESS: *Given any numbers at all, one will list none of the others.*

Proof Say that some numbers are, collectively, FRIENDLY if each of them lists at least one of the others. Suppose you are presented with some friendly numbers. Say that a number is UNFRIENDLY if it lists none of those friendly numbers. Since ∅ lists nothing, it is unfriendly. Suppose n is unfriendly. If one of n's entries listed one of the friendly numbers, then, by Theorem 3.6, n would do so too. So n's entries are all unfriendly. But then, by Theorem 3.7, Sn lists none of the friendly numbers and, so, is itself unfriendly. We conclude, by induction, that there really are no friendly numbers. □

Theorem 3.9 CONNECTEDNESS: *Given any two numbers, one will list the other.*

Proof Say that two numbers are CONNECTED if and only if one of them lists the other. In Exercise 3.19, you showed that ∅ is connected with every other number. Suppose n is so as well. Let k be a number distinct from and unconnected with Sn. k either lists or is listed by n. It cannot be the latter, since k would then be listed by Sn. So k lists n and, hence, (as you showed in Exercise 3.18) Sk lists Sn. This means that Sn is either k (contrary to our assumption that they are distinct) or an entry of k (contrary to our assumption that they are unconnected). We conclude, by induction, that every number is connected with every other number. □

---

[11] See Hallett [6], pp. 185–194, and Mirimanoff [10].

Since the relation of listing, when applied to numbers, is transitive, irreflexive, well-founded, and connected, it is said to WELL-ORDER the numbers.

Exercise 3.20 *Show that if m and n are numbers, then there is a list $m \backslash n$ that lists, exactly, the numbers that m lists but n does not. What is $13 \backslash 9$? What is $9 \backslash 13$?*

3.7 Super-Numbers

Recall that if x is a list, then Px is the list of x's parts.

Definition 3.8 Some lists are, collectively, P-CLOSED if and only if the list Px is one of them whenever the list x is.

Definition 3.9 A list x is a SUPER- NUMBER if and only if it satisfies the following condition: given any P-closed lists, if $\emptyset$ is one of them, then so is x.

Super-numbers are what tireless immortal beings encounter when they start with $\emptyset$ and persist in applying the operation P to everything they encounter. The first few super-numbers are $\emptyset$, $P\emptyset$, $PP\emptyset$, $PPP\emptyset$. That is:

$$\emptyset$$

$$\{\emptyset\}$$

$$\{\emptyset, \{\emptyset\}\}$$

$$\{\emptyset, \{\emptyset\}, \{\{\emptyset\}\}, \{\emptyset, \{\emptyset\}\}\}$$

Theorem 3.10 $\emptyset$ *is a super-number.*

Theorem 3.11 *If x is a super-number, then so is Px.*

Super-numbers behave very much like numbers (hence the name).[12]

Theorem 3.12 $Py \neq \emptyset$.

Exercise 3.21 *Show that $Px = Py$ only if $x = y$.*

Theorem 3.13 *If some lists are P-closed and $\emptyset$ is one of them, then every super-number is one of them.*

Note that the last theorem is a kind of induction principle. You can show that every super-number has a certain property by showing that $\emptyset$ has the property and that each super-number x passes the property on to its "successor" Px.

---

[12] I should warn you, though, that the term 'super-number' is not standard. Do not expect anyone else to refer to these objects in this way.

Exercise 3.22 *Show that every super-number is pure.*

Exercise 3.23 *Show that every super-number lists the parts of its entries.*

Exercise 3.24 *Show that if super-number x lists y, then Px lists Py.*

Theorem 3.14 *If x and y are super-numbers and x is a proper part of y, then x is an entry of y.*

Proof The theorem comes out true when $x = \emptyset$ because every super-number other than $\emptyset$ lists $\emptyset$. (Every super-number lists the parts of its entries and $\emptyset$ is a part of every entry.) Suppose, as an inductive hypothesis, that the super-number x is an entry of a super-number whenever it is a proper part. Suppose Px is a proper part of the super-number y. Then $y \neq \emptyset$ and we can show that $y = Py'$ for some super-number y'. (Go ahead and prove it if you do not already believe it.) So Px is a proper part of Py' and, as you showed in Exercise 3.12, x is a proper part of y'. Our inductive hypothesis now guarantees that x is an entry of y'. So, by Exercise 3.23, y' lists all the parts of x and, hence, Px is a part of y'. That is, Px is an entry of Py', as desired. We have shown that $\emptyset$ satisfies the theorem and that the super-numbers satisfying the theorem are, collectively, P-closed. So every super-number satisfies the theorem. You will probably find this method of induction useful. (Perhaps you already found it useful when you did Exercises 3.22 and 3.23.) □

Exercise 3.25 *Show that every super-number lists the entries of its entries (so every entry of a super-number x is a part of x).*

Exercise 3.7 now allows us to prove the following.

Theorem 3.15 *The precursors of a super-number x are x itself and x's entries.*

Exercise 3.26 *Show that, given any two super-numbers, one will list the other. (As in the proof of Theorem 3.9, say that two super-numbers are* CONNECTED *if and only if one of them lists the other. As an inductive hypothesis, suppose x is a super-number connected with every other super-number. Let y be a super-number distinct from Px. Note that either x lists y or y lists x. To confirm that y and Px are connected, consider each of these cases.)*

Definition 3.10 A list is WELL-FOUNDED if and only if, given any of its precursors, at least one of them lists none of them.

Earlier in this chapter, we considered a list that may have struck you as odd. Here it is again.

My list of three piano sonatas

My list of three phrases

My list of three lists

My list of three lists was this very list. If called upon to pick some precursors of this list, we could, if we wished, pick the list itself and then decline to pick anything else. Then all of the precursors we have picked (all one of them) list a precursor we have picked. So the above list is not well-founded: we can find an endless path through its precursors that takes us from a precursor (my lists of three lists) to an entry of that precursor (my lists of three lists) to an entry of *that* precursor (my lists of three lists) and so on forever. Quite generally, lists that list themselves

$$x \rightsquigarrow x$$

are not well-founded. Neither are lists listed by something they list

$$y \rightsquigarrow z \rightsquigarrow y.$$

Nor is a list with infinitely many precursors $w_1, w_2, w_3, \ldots$ each listing the next

$$w_1 \rightsquigarrow w_2 \rightsquigarrow w_3 \rightsquigarrow \cdots$$

In each case, we have precursors each of which lists at least one of those very precursors. (x lists x. y lists z, while z lists y. w_n lists w_{n+1}.)

Theorem 3.16 *Every super-number is well-founded*

Proof The only precursor of $\emptyset$ (itself) lists nothing. So $\emptyset$ is well-founded. Suppose x is a well-founded super-number. Let $y_1, y_2, y_3, \ldots$ be precursors of Px that violate well-foundedness. (That is, each y_k lists some y_j.) Theorems 3.11 and 3.15 guarantee that each y_k is either Px itself or an entry of Px. Each case yields a y_j that is an entry of x. (Suppose, first, that y_k is an entry of Px. Let y_j be an entry of y_k. Then y_j is an entry of x since it is an entry of a part of x. Now suppose $y_k = Px$. Let y_i be an entry of y_k. y_i is a part of x. Pick a y_j listed by y_i. y_j is an entry of x.) Let $w_1, w_2, w_3, \ldots$ be the y_j's listed by x. Pick a w_k. Since w_k appears among $y_1, y_2, y_3, \ldots$ and these lists violate well-foundedness, we can pick a y_j listed by w_k. By Exercise 3.25, x lists y_j and, hence, y_j appears among $w_1, w_2, w_3, \ldots$. More generally, each w_k lists some y_j that appears among $w_1, w_2, w_3, \ldots$. That is, each of the lists $w_1, w_2, w_3, \ldots$ lists at least one of the lists $w_1, w_2, w_3, \ldots$. Since $w_1, w_2, w_3, \ldots$ are all precursors of x, this contradicts the well-foundedness of x. So we must have been wrong to deny that Px is well-founded. Since $\emptyset$ is well-founded and the well-founded super-numbers are, collectively, P-closed, every super-number is well-founded. □

3.8 Hereditary Finiteness

A finite list is a list with finitely many entries. *Hereditarily* finite lists are finite "all the way down." Their entries have finitely many entries, as do the entries of their entries, the entries of the entries of their entries, and so on. The pure, well-founded, hereditarily finite lists are the lists coded by numbers in the Ackermann interpretation. They are also the lists listed by super-numbers. Now "pure, well-founded, hereditarily finite" is quite a mouthful. We will follow set theoretic practice and shorten it to just "hereditarily finite" even though the odd list we discussed in the previous section is hereditarily finite without being either pure or well-founded. Whenever you see "hereditarily finite," try to remember that there are invisible qualifications: "pure and well-founded."

Definition 3.11 A list is hereditarily finite if and only if it is listed by a super-number.

Exercise 3.27 *Show that each number is hereditarily finite. (After confirming that a super-number lists Ø, you might assume, as an inductive hypothesis, that a super-number x lists the number n. Theorem 3.7 might help you show that Sn is a part of x and, hence, that Px lists Sn.)*

Definition 3.12 If x is pure, then $\bigcup x$ is the list of the entries of x's entries.

It may not be obvious why Definition 3.12 requires that x be pure (that is, that x's precursors all be lists). Consider the impure list {{Frege}, {Hilbert}}. This list has two entries: {Frege} and {Hilbert}. The first of these entries has a single entry: Frege. The second also has a single entry: Hilbert. So Frege and Hilbert are the entries of the entries of {{Frege}, {Hilbert}}. Now there is nothing to prevent Frege and Hilbert from forming a list: {Frege, Hilbert}. This would be the list $\bigcup${{Frege}, {Hilbert}} or, in notation that may be more familiar, {Frege} $\cup$ {Hilbert}. Remember, though, what our current project is. We are working inside an axiomatic system. We are trying to see what follows from Axioms 3.1–3.3. No matter how obvious it may seem that a certain list exists, we cannot accept that it exists unless our axioms confirm this. The relevant axiom here is 3.3:

> Any lists of precursors of a list form a list.

Frege and Hilbert are, indeed, precursors of {{Frege}, {Hilbert}}, but they are not lists of precursors because (presumably) they are not lists. So Axiom 3.3 does not apply. This is why Theorem 3.1

> Any precursors of a pure list form a list

includes the qualification "pure." Consider, for example, the pure list {{Ø}, {{Ø}}}. It has two entries: {Ø} and {{Ø}}. The first of these entries has a single entry: Ø. The second also has a single entry: {Ø}. So Ø and {Ø} are the entries of the entries of {{Ø}, {{Ø}}}. Furthermore, they are lists of precursors of {{Ø}, {{Ø}}} and, so, Axiom 3.3 assures us that they form a list. That list, {Ø, {Ø}}, is $\bigcup${{Ø}, {{Ø}}} or, in slightly different notation, {Ø}$\cup${{Ø}}. More generally, if a list is pure, then Axiom 3.3 assures us that the entries of its entries really do form a list. If a list is impure, we receive no such assurance.

Exercise 3.28 *Show that if x is hereditarily finite, then so is $\bigcup x$. (You might assume that a super-number y lists x and then try to show that $\bigcup x$ is a part of y.)*

Exercise 3.29 *Show that if x is hereditarily finite, then so is Px.*

Exercise 3.30 *Show that if x and y are hereditarily finite, then $\{x, y\}$ exists and is hereditarily finite. (Exercise 3.26 might be useful.)*

Exercise 3.31 *Show that if some lists are all entries of a hereditarily finite list, then they form a hereditarily finite list.*

Exercise 3.32 *Show that hereditarily finite lists with the same hereditarily finite entries are the same.*

Exercises 3.28–3.32 show that five of the axioms from the theory Z, Zermelo's classic axiomatization of set theory, can be interpreted as statements about hereditarily finite lists: just read "set" as "hereditarily finite list" and "is an element of" as "is an entry of."[13]

The following theorem will be useful in Chaps. 4 and 5.

Theorem 3.17 *No hereditarily finite list lists every number.*

Proof By Exercise 3.25, if some hereditarily finite list listed every number, some super-number would do so as well. We will use Theorem 3.13 (super-number induction) to show that this is impossible. Note, first, that $\emptyset$ omits every number. Suppose x omits the number n. Then Sn is not a part of x since Sn lists n. So Px omits Sn.

$\square$

3.9 Finite Ranks

Here is a fairy tale. In spite of their name, the Listers have never done any listing. Now they are going to make lists galore. They proceed systematically, in a series of stages. At each stage, they list lists from earlier stages. They are careful not to reproduce earlier lists; but they are also careful not to miss any opportunity to make new lists. The Listers will list any earlier lists as long as the result is a new list.

The process begins at stage 0. Since the Listers have never listed, their first list of earlier lists is blank.

Stage 0 yields $\emptyset$.

At the next stage, stage 1, the Listers list the one and only earlier list. Their one new list is the list whose only entry is the blank list.

---

[13] See Zermelo [14]; English translation in van Heijenoort [13], pp. 199–215.

Stage 1 yields {Ø}.

When they reach stage 2, the Listers have already made two lists: Ø and {Ø}.[14] They obtain one new list by listing both of these lists. They obtain another new list by listing {Ø} all by itself. They do not list Ø all by itself because they already did that at stage 1.

Stage 2 yields {{Ø}} and {Ø, {Ø}}.

They now proceed to stages 3, 4, 5 and beyond.

Exercise 3.33 *Define a function that gives the number of lists appearing at stage n. (You can use recursion if you want.) Do you think the Listers can complete stage 4? Do you think they can really complete stage 5?*

If a list appears at stage n in the Listers' listing process, we say that its RANK is n. Ø has rank 0. {Ø} has rank 1. {{Ø}} and {Ø, {Ø}} have rank 2. A graph may help to make the notion of rank more vivid. Here are the Listers' first four lists with arrows representing the relation "is an entry of."

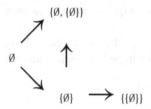

Note that the rank of list x is the number of arrows in the longest path from Ø to x. In the Listers' listing process, a list appears just after all its entries have appeared. Each list has to wait for the appearance of its latest arriving entries. When we count arrows in the *longest* path from Ø to x, we are letting the ranks of x's latest arriving entries determine the rank of x.

Exercise 3.34 *Determine the rank of {{Ø}, {{{Ø}}}}. Determine the rank of*

$$\{\{\{\{\{\{\{\{\{Ø\}\}\}\}\}\}\}\}\}.$$

How does the bracket notation "{...}" help us determine ranks?

Here is yet another way to think about ranks. In a moment, I will introduce some lists

---

[14] OK, I am being sloppy here. Stated more precisely, the Listers have made tokens of the two types Ø and {Ø}. We do not need to imagine that they *make* the types. At each stage, they list types that have tokens produced at earlier stages.

$$V(0), V(1), V(2), \ldots$$

that will turn out to be our old friends the super-numbers

$$\emptyset, P\emptyset, PP\emptyset, \ldots.$$

Here are $V(0), V(1), V(2), V(3)$.

$$\emptyset$$

$$\{\emptyset\}$$

$$\{\emptyset, \{\emptyset\}\}$$

$$\{\emptyset, \{\emptyset\}, \{\{\emptyset\}\}, \{\emptyset, \{\emptyset\}\}\}$$

Although $V(0)$ lists nothing, each subsequent $V(n)$ lists the lists of rank less than n. $V(1)$ lists the one list of rank 0. $V(2)$ lists the one list of rank 0 and the one list of rank 1. $V(3)$ lists the one list of rank 0, the one list of rank 1, and the two lists of rank 2. Lists that first appear on list $V(n+1)$ have rank n. A list appears on $V(n+1)$ if and only if it is part of $V(n)$. So a list of rank n will be a part of $V(n)$, but will not be a part of $V(k)$ for any k less than n. This will motivate the definition of rank that appears below.

Though I will not give the details, we can use induction and some other tricks to show that if n is a number, then there is a unique sequence of lists $a_0, a_1, \ldots, a_n$ with the following properties.[15]

$$a_0 = \emptyset$$

$$a_{Sk} = Pa_k$$

If we identify $V(n)$ with the last term a_n in the sequence $a_0, a_1, \ldots, a_n$, we can confirm that the lists $V(0), V(1), \ldots$ have the following properties.

$$V(0) = \emptyset$$

$$V(Sk) = PV(k)$$

Assume that this is so. (Or verify it yourself, if you want.) You can now show that the lists $V(n)$ are, in fact, our super-numbers.

Exercise 3.35 *Use induction to show that if n is a number, then $V(n)$ is a super-number.*

---

[15] Among the details I am skipping over is the question of whether Axioms 3.1–3.3 allow us to introduce the notion of finite sequence. You might find it interesting to show that they do.

Exercise 3.36 *Use Theorem 3.13 (i.e., super-number induction) to show that if x is a super-number, then $x = V(n)$ for some number n.*

Now that you have confirmed that the lists $V(0), V(1), V(2), \ldots$ are the super-numbers $\emptyset, P\emptyset, PP\emptyset, \ldots$, you can use facts about super-numbers to obtain results about the $V(n)$'s. For example, you might want to use Exercise 3.25 to complete the following exercise.

Exercise 3.37 *Use induction (on n) to show that if n lists m, then $V(n)$ lists $V(m)$.*

Theorem 3.18 *If $V(m) = V(n)$, then $m = n$.*

Proof Suppose $V(m) = V(n)$. According to Theorem 3.9, if m and n are distinct, then one lists the other. Suppose n lists m. Then, by the result you just established, $V(n)$ lists $V(m)$ and, hence, $V(n)$ lists $V(n)$. Since this contradicts Theorem 3.16, we conclude that m and n are not distinct. $\square$

Exercise 3.38 *Show that if $V(n)$ lists $V(m)$, then n lists m. (You might find Theorem 3.16 useful.)*

Since the hereditarily finite lists are the entries of the super-numbers, every hereditarily finite list is an entry of $V(n)$ for some number n. By Exercise 3.25, every entry of $V(n)$ is a part of $V(n)$. Theorem 3.8 assures us that if x is a part of $V(n)$ for some number n, then there is a smallest such n. So the following definition is justified.

Definition 3.13 If x is hereditarily finite, then $\rho(x)$ (the RANK of x) is the first number n such that x is a part of $V(n)$.

For the rest of this section, we will assume that x, y, z are hereditarily finite lists.

Exercise 3.39 *Use induction on n to show that if $V(n)$ lists x, then n lists $\rho(x)$. (Note that if x is a part of $V(n)$, then $\rho(x)$ is no greater than n: that is, either n lists $\rho(x)$ or $n = \rho(x)$.)*

Definition 3.14 List y OUTRANKS list x if and only if $\rho(y)$ lists $\rho(x)$.

Exercise 3.40 *Show that every hereditarily finite list outranks its entries.*

Exercise 3.41 *Show that if z outranks each entry of y, then y does not outrank z. (You might find Exercise 3.23 useful.)*

Exercise 3.41 says that each hereditarily finite list has the lowest rank compatible with Exercise 3.40.

Exercise 3.42 *Show that if y does not outrank z, then z outranks each entry of y. (Note: since our ranks are numbers, they obey Theorem 3.9.)*

Theorem 3.19 *One hereditarily finite list outranks another if and only if some hereditarily finite entry of the former outranks each hereditarily finite entry of the latter.*

Proof According to Exercises 3.41 and 3.42, y fails to outrank z if and only if z outranks each entry of y. This is equivalent to:

$$y \text{ outranks } z \iff z \text{ fails to outrank some entry of } y.$$

Applying Exercises 3.41 and 3.42 again, z fails to outrank a list x if and only if x outranks each entry of z. So z fails to outrank some entry of y if and only if some entry of y outranks each entry of z. Our grand conclusion:

$$y \text{ outranks } z \iff \text{ some entry of } y \text{ outranks each entry of } z.$$

Exercise 3.25 guarantees that each entry of a hereditarily finite list is hereditarily finite. So, given our standing assumption that y is hereditarily finite, "some entry of y" is necessarily "some hereditarily finite entry of y."   □

Definition 3.15 Some hereditarily finite lists are, collectively, of BOUNDED RANK if and only if some hereditarily finite list outranks all of them.

Exercise 3.43 *Show that any hereditarily finite lists of bounded rank will form a hereditarily finite list.*

3.10 And Why Are We Doing This?

Although we are marching under the banner of list theory, this is just a way of talking about set theory. So you have already received a large dose of set theory. You will receive two more doses in Chaps. 4 and 5. Perhaps, then, I should pause and reassure you that this treatment really is warranted.

This book is supposed to provide a mathematical prelude to the philosophy of mathematics. It is supposed to help you understand philosophical work that has already been done and is also meant to prepare you to do philosophical work of your own. A big dose of set theory should help you both understand and do philosophy of mathematics. First of all, many philosophers of note have a lot to say about set theory. You will want to understand what they are saying. Secondly, if *you* are to have any chance of saying something intelligent about contemporary mathematics, you will need to have a good idea of what contemporary mathematics is. No whirlwind tour of the contemporary scene is better than the one provided by set theory. Even if you become a professional mathematician, all the mathematics you are likely to encounter or produce will be reducible to the set theories presented in Chaps. 3, 4 and 5 or to some natural extension of those set theories. Granted, these three chapters will not magically introduce you to all the specialized vocabularies, techniques, and results of all the sub-fields of mathematics.[16] Your apprenticeship in set theory will, however, give you a general idea of what counts as a possible structure or operation in today's mathematics. That is a goal worth pursuing. We continue our pursuit of it in the next chapter.

3.11 Solutions of Odd-Numbered Exercises

3.1 Let x be the list of all lists that do not list themselves. Does x list itself? It does so only if it is one of the lists that do not list themselves. That is, it does only if it does not. So x does not list itself and, so, appears on the list of all lists that do not list themselves. But that list is x itself. We conclude that x both does and does not list itself. That is, x *would* do this impossible thing if it existed. So no such list exists.

3.3 We are going to treat "forming a pedigree" as a property that people possess collectively. Some people FORM A PEDIGREE if and only if each parent of one of them is one of them. x is an ANCESTOR of y if and only if, given any people who form a pedigree, if y is one of those people, then so is x.

3.5 Suppose w is a precursor of x and x is a precursor of y. Given any entry-closed lists, if y is one of them, so is x. Given any entry-closed lists, if x is one of them, so is w. So, given any entry-closed lists, if y is one of them, so is w. That is, w is a precursor of y.

3.7 Let "the Y-lists" be y itself and y's entries. Suppose w is an entry of x and x is one of the Y-lists. If $x = y$, then w is one of the Y-lists by virtue of being an entry of y. Suppose x is an entry of y. Then w is an entry of an entry of y and, hence, is an entry of y. So, once again, w is one of the Y-lists. We conclude that the Y-lists are entry-closed. Suppose z is a precursor of y. Given any entry-closed lists, if y is one of them, so is z. So z is one of the Y-lists. That is, z is either y or an entry of y.

3.9 Suppose x is a precursor of Py. By Exercise 3.6, x is either Py (and, hence, a list) or a precursor of an entry of Py. Suppose the latter. Then x is a precursor of a part of y. By Exercise 3.6 again, x is either a part of y (and, hence, a list) or a precursor of an entry of a part of y. Suppose the latter. Then x is a precursor of an entry of y and, hence, by Exercise 3.5, is a precursor of y. If y is pure, x is a list. We conclude that all of Py's precursors are lists.

3.11 Since y is one of its own precursors, Sy lists y. $\emptyset$ lists nothing.

---

[16] To see how unlikely that is, take a look at the "Mathematics Subject Classification" at http://www.ams.org/msc/pdfs/classifications2010.pdf. Note that this document is 47 pages long.

3.13 18 is 10010 ($1 \cdot 2^4 + 0 \cdot 2^3 + 0 \cdot 2^2 + 1 \cdot 2^1 + 0 \cdot 2^0$) in binary. Here is the decoding key.

List 100	List 11	List 10	List 1	List 0
1	0	0	1	0

We already know that List 1 is {Ø}. We need to figure out what List 100 is. Here is the decoding key for that.

List 10	List 1	List 0
1	0	0

We know that List 10 is {{Ø}}. So List 100 is {{{Ø}}} since it says yes only to List 10. List 10010 says yes only to Lists 100 and 1. So it is {{{{Ø}}}, {Ø}}.

3.15 Given any S-closed lists, if Ø is one of them, then Ø is one of them.

3.17 Suppose $Sm = Sn$. Since Sm lists m, so does Sn. By Theorem 3.7, m is either n or an entry of n. Since Sn lists n, so does Sm. By Theorem 3.7, n is either m or an entry of m. Suppose $m \neq n$. Then m is an entry of an entry of m. So, by Theorem 3.6, m is an entry of itself—contrary to Exercise 3.16.

3.19 We want to show that every number n has the following property: if $n \neq$ Ø, then n lists Ø. It is vacuously true that Ø has this property. As an inductive hypothesis, suppose n has the property. That is, n either is Ø or lists Ø. So Ø is either an entry or an entry of an entry of Sn. By Theorem 3.6, Ø is an entry of Sn.

3.21 Suppose $Px = Py$. Since Px lists x, so does Py. That is, x is a part of y. Since Py lists y, so does Px. That is, y is a part of x. So x and y have the same entries and, hence, by Axiom 3.1, $x = y$.

3.23 It is vacuously true that Ø lists the parts of its entries. Suppose y is a part of an entry of Px. Then y is a part of a part of x and, hence, is a part of x. So Px lists y. This is one of those odd inductions without an inductive hypothesis.

3.25 It is vacuously true that Ø lists the entries of its entries. Suppose, as an inductive hypothesis, that x lists the entries of its entries. Suppose y is an entry of an entry of Px. Then y is an entry of a part of x and, hence, is an entry of x. By our inductive hypothesis, y's entries are entries of x and, hence, y is a part of x. So y is an entry of Px.

3.27 By Theorems 3.10 and 3.11, $P\emptyset$ is a super-number. So $\emptyset$ is hereditarily finite since $P\emptyset$ lists $\emptyset$. Suppose, as an inductive hypothesis, that n is hereditarily finite. Pick a super-number x that lists n. Suppose Sn lists y. Then, by Theorem 3.7, y is either n or an entry of n. That is, y is either an entry of x or an entry of an entry of x. By Exercise 3.25, y is an entry of x. We conclude that Sn is a part of x and, hence, is an entry of Px. This confirms that Sn is hereditarily finite if n is.

3.29 By Theorem 3.11 and Exercise 3.24, if super-number y lists x, then super-number Py lists Px.

3.31 By Exercise 3.25, any entries of a hereditarily finite list are entries of a super-number. By Theorem 3.2 and Exercise 3.22, any entries of a super-number form a list. That list will be part of the super-number. So, if the super-number is x, the list will be an entry of Px.

3.33 Suppose you are making a list and there are n things that could appear as entries. You are free to list all, some, or none of those n things. Once you finish your list, each of the n things will be in one of two possible states: on the list or not. Your list will record n decisions about two possible outcomes. So there are 2^n lists you could make. If you start with 0 things, the number of possible lists is 2^0: you can only produce the blank list $\emptyset$. Repeat the process starting with the one blank list and the number of possible lists is 2^1: you could produce either $\emptyset$ or $\{\emptyset\}$. Repeat the process starting with these two lists and the number of possible lists is 2^2. The number of lists that could appear at each stage of this process is given by the following function.

$$a(0) = 0$$
$$a(n+1) = 2^{a(n)}$$

There is, however, a complication. The Listers do not reproduce lists that have already appeared. The following function makes the necessary correction.

$$f(n) = a(n+1) - a(n)$$

To complete stage 4, the Listers would have to list 65,520 lists. If they listed one list per second, this would take 18 h and 12 min. To complete stage 5, the Listers would have to list $2^{65,536} - 65,536$ lists. At one list per second, this would take about $10^{19,775}$ times the age of the universe.

3.35 By Theorem 3.10, $V(0)$ is a super-number. By Theorem 3.11, $PV(n)$ is a super-number if $V(n)$ is. So $V(Sn)$ is a super-number if $V(n)$ is.

3.37 It is vacuously true that the result holds when $n = \emptyset$. As an inductive hypothesis, suppose n lists m only if $V(n)$ lists $V(m)$. Suppose Sn lists m. Since $V(Sn) = PV(n)$, we want to show that $V(m)$ is a part of $V(n)$ (and, hence, an entry of $PV(n)$). By Theorem 3.7, m is either n or an entry of n. If $m = n$, then $V(m)$ is a part of $V(n)$ since every list is a part of itself. Suppose m is an entry of n. Then, by our inductive hypothesis, $V(n)$ lists $V(m)$ and, hence, by Exercise 3.25, $V(m)$ is a part of $V(n)$.

3.39 It is vacuously true that the result holds when $n = \emptyset$. Suppose $V(Sn)$ lists x. Then x is an entry of $PV(n)$ and, hence, a part of $V(n)$. $\rho(x)$ is the *first* number that behaves like this n. So $\rho(x)$ is either n or an entry of n. By Theorem 3.7, $\rho(x)$ is an entry of Sn.

3.41 Suppose z outranks each entry of y. That is, $\rho(z)$ lists $\rho(x)$ whenever y lists x. By Exercise 3.37, $V(\rho(z))$ lists $V(\rho(x))$ whenever y lists x. x is a part of $V(\rho(x))$. So, by Exercise 3.23, $V(\rho(z))$ lists x whenever y lists x. That is, y is a part of $V(\rho(z))$. $\rho(y)$ is the first number n such that y is a part of $V(n)$. So $\rho(z)$ cannot be smaller than $\rho(y)$.

3.43 Pick a hereditarily finite list y and some hereditarily finite lists "the X's" with the following property: if x is one of the X's, then $\rho(y)$ lists $\rho(x)$. By Exercise 3.37, if x is one of the X's, then $V(\rho(y))$ lists $V(\rho(x))$. Every hereditarily finite x is part of $V(\rho(x))$. So, by Exercise 3.23, if x is one of the X's, then $V(\rho(y))$ lists x. We conclude that the X's are all entries of $V(\rho(y))$. Theorem 3.2 and Exercise 3.22 now yield a list whose entries are exactly the X's. Since that list is a part of $V(\rho(y))$, it is an entry of $PV(\rho(y))$ and, so, is hereditarily finite.

References

1. Ackermann, W. (1937). Die Widerspruchsfreiheit der allgemeinen Mengenlehre. *Mathematische Annalen, 114*, 305–315.
2. Bennett, D. (2000). A single axiom for set theory. *Notre Dame Journal of Formal Logic, 41*, 152–170.
3. Deiser, O. (2011). An axiomatic theory of well-orderings. *Review of Symbolic Logic, 4*, 186–204.
4. Edwards, H. M. (1977). *Fermat's last theorem: A genetic introduction to algebraic number theory*. New York: Springer-Verlag.
5. Frege, G. (1879). *Begriffsschrift*. Halle: Louis Nebert.
6. Hallett, M. (1984). *Cantorian set theory and limitation of size*. Oxford: Clarendon Press.
7. Kanamori, A. (2004). Zermelo and set theory. *Bulletin of Symbolic Logic, 10*, 487–553.
8. Kaye, R., & Wong, T. L. (2007). On interpretations of arithmetic and set theory. *Notre Dame Journal of Formal Logic, 48*, 497–510.
9. Landini, G. (2013). Zermelo and Russell's paradox: Is there a universal set? *Philosophia Mathematica, 21*, 180–199.
10. Mirimanoff, D. (1917). Les antinomies de Russell et de Burali-Forti et le problème fondamental de la théorie des ensembles. *L'Enseignement Mathématique, 19*, 37–52.

11. Pollard, S. (1990). *Philosophical introduction to set theory.* Notre Dame IN: University of Notre Dame Press.

12. Rang, B., & Thomas, W. (1981). Zermelo's discovery of the 'Russell paradox'. *Historia Mathematica, 8,* 15–22.

13. van Heijenoort, J. (Ed.). (1967). *From Frege to Gödel.* Cambridge MA: Harvard University Press.

14. Zermelo, E. (1908). Untersuchungen über die Grundlagen der Mengenlehre I. *Mathematische Annalen, 65,* 261–281.

Chapter 4
Zermelian Lists

4.1 Infinite Lists

Given any natural number n, our principles guarantee us a list with n entries. Indeed, in our construction, the number n is itself a list with n entries. 10^{100} is a list with 10^{100} entries. Now we human beings are never going to list 10^{100} things. We might *describe* them in some way; but we normally think of a description as an *alternative* to a list, not a *kind* of list. So our principles posit a world inhabited by infinitely many lists, infinitely many of which surpass, to an extravagant degree, our limited capacity to list things. We may not believe there is such a world; but the Ackermann interpretation discussed in the last chapter provides an excellent reason to believe there could be. As mathematicians, we have found it interesting and productive to explore what such worlds would be like.

Having already left far behind what is humanly feasible in our pursuit of what is logically possible, we may feel comfortable with another leap. We already have infinitely many lists. How about a list with infinitely many entries? What might the entries be? Well, there are infinitely many hereditarily finite sets. Perhaps there could be a list of them. Let us try out that idea.

Axiom 4.1 *The hereditarily finite lists form a list.*

The list of all hereditarily finite lists is known as $V(\omega)$. ('ω' is the Greek letter omega.) As before, we claim no certain knowledge that our axiom is true, that there is such a thing as $V(\omega)$. We are just letting everyone know we are interested in what the world would be like if there were.

Exercise 4.1 *Show that $V(\omega)$ is pure.*

Exercise 4.2 *Show that the numbers form a list.*

The list of all numbers is known as 'ω'. This may help to explain the name '$V(\omega)$'. If we start with $\emptyset$, then n applications of the operation P will give us $V(n)$. If we start with $\emptyset$, then ω applications of the operation P will give us $V(\omega)$. In conventional set theoretic terminology:

$$V(0) = \emptyset$$

$$V(n + 1) = PV(n)$$

$$V(\omega) = \bigcup_{n \in \omega} V(n)$$

Something is listed by $V(\omega)$ if and only if it is listed by $V(n)$ for some number n. You proved in Chap. 3 that the $V(n)$'s are the super-numbers. So $V(\omega)$ is what you get when you list everything that appears in the construction of the super-numbers.

$$V(\omega) = V(0) \cup V(1) \cup V(2) \cup \cdots = \emptyset \cup P\emptyset \cup PP\emptyset \cup \cdots$$

$V(\omega)$ lists everything you obtain when you start with $\emptyset$ and complete ω applications of the operation P, one application for each number $0, 1, 2, 3, \ldots$.

We need to consider whether our new system, consisting of Axioms 3.1, 3.2, 3.3, and now 4.1, is consistent. First, though, another question. Is it really helpful, does it really aid understanding, to continue to use the word "list" at this point? Hasn't our everyday notion of list been stretched beyond recognition when we start talking about infinite lists? Better, perhaps, to avoid unnecessary confusion and resistance by saying good-bye to "list" and introducing a new term not so loaded down with preconceptions.

Well, I am not sure one person can speak for everyone on the question of how flexible our concepts are. My concept of list might be far more flexible than yours. How then would we decide whether the concept has been stretched too far? Luckily, we do not need to reach any such decision. We already know that we will need to abandon "list" before long. We want to be understood by other English-speaking mathematicians; and they say "set" not "list." We need to learn to say "set" too. We shall stick to "list" for just a few more pages.

Now, is our new system consistent? It turns out to be impossible to interpret all four of our axioms as truths of arithmetic. Ackermann's trick cannot help us here. But we still have good evidence that our system is consistent. Zermelo managed to describe a structure, $V(\omega + \omega)$, in which all our principles would be true.[1] Some of the most profound mathematicians of the last century thought deeply about that structure, gave strong indications of having formed a clear concept of it, and reported no signs of incoherence. That is not a *proof* of consistency. But proof is not the only kind of evidence available to mathematicians nor is it always the best kind of evidence. Our concern is that we might spend a lot of time seeing what follows from premises that turn out to be absurd. Since everything would be true in a logically impossible world, the painstaking accumulation of theorems is quite misguided when the axioms are incoherent. Ours might be; but we have very good reason to think they are not. Let us be bold and see what sort of world we are now contemplating.

You might think of $\omega + \omega$ as the result of the following construction.

---

[1] See Zermelo [2].

$$\omega + 0 = \omega$$

$$\omega + Sn = S(\omega + n)$$

$$\omega + \omega = \bigcup_{n \in \omega} (\omega + n)$$

That is,

$$\omega + \omega = (\omega + 0) \cup (\omega + 1) \cup (\omega + 2) \cup \cdots = \omega \cup S\omega \cup SS\omega \cup \cdots$$

$\omega + \omega$ lists everything you obtain when you start with ω and complete ω iterations of the operation S.

You can think of $V(\omega + \omega)$ as the result of the following construction.

$$V(\omega + 0) = V(\omega)$$

$$V(\omega + Sn) = PV(\omega + n)$$

$$V(\omega + \omega) = \bigcup_{n \in \omega} V(\omega + n)$$

That is,

$$V(\omega+\omega) = V(\omega+0) \cup V(\omega+1) \cup V(\omega+2) \cup \cdots = V(\omega) \cup PV(\omega) \cup PPV(\omega) \cup \cdots$$

$V(\omega + \omega)$ lists everything you obtain when you start with $V(\omega)$ and complete ω iterations of the operation P.

Our four axioms allow us to show that $\omega + n$ and $V(\omega + n)$ exist for each number n. They do not allow us to show that either $\omega + \omega$ or $V(\omega + \omega)$ exist. Our four axioms and the axioms of Zermelo's set theory Z can all be interpreted as statements about the entries of $V(\omega + \omega)$. That is, to make our axioms and Zermelo's axioms true, you need only (1) complete ω applications of the operation P starting from the blank list; (2) list every list you encounter in this construction; then (3) starting from that list, complete ω additional applications of P.

Exercise 4.3 *Dream up an axiom that guarantees the existence of both $\omega + \omega$ and $V(\omega + \omega)$.*

Exercise 4.4 *Suppose x and y are parts of $V(\omega)$. Show that there is a list $x \cup y$ that lists, exactly, the entries of x and the entries of y.*

Exercise 4.5 *Show that ω lists the entries of its entries or, in other words, that every number is a list of numbers.*

Exercise 4.6 *Suppose x is a part of ω. Show that there is a list $\omega \setminus x$ that lists, exactly, the numbers that x does not list.*

4.2 More Ranks

We now continue the discussion of ranks we began in the previous chapter. We will assign ranks to lists in our new, larger universe. We can use induction to show that if n is a number, then there is a unique sequence of lists $a_0, a_1, \ldots, a_n$ and a unique sequence of lists $b_0, b_1, \ldots, b_n$ with the following properties.

$$a_0 = \omega$$

$$a_{Sk} = Sa_k$$

$$b_0 = V(\omega)$$

$$b_{Sk} = Pb_k$$

If we identify $\omega + n$ with the last term a_n in the sequence $a_0, a_1, \ldots, a_n$ and identify $V(\omega + n)$ with the last term b_n in the sequence $b_0, b_1, \ldots, b_n$, we can confirm that the lists $\omega + 0, \omega + 1, \ldots$ and $V(\omega + 0), V(\omega + 1), \ldots$ have the properties we noted above:

$$\omega + 0 = \omega$$

$$\omega + Sn = S(\omega + n)$$

$$V(S(\omega + n)) = PV(\omega + n).$$

Assume this is so. (Or verify it yourself, if you want.) You can now show that the lists $\omega + n$ are the lists we are going to call HIGHER RANKS:

$$\omega, S\omega, SS\omega \ldots.$$

The higher ranks are what tireless immortal beings encounter when they start with ω (the list of all numbers) and apply the operation S to everything they encounter.

Exercise 4.7 *Using* Definition 3.7 *as your guide, say what it means for something to be one of the higher ranks ω, $S\omega$, $SS\omega \ldots$.. Confirm that the higher ranks are the lists $\omega + n$.*

You can also show that the lists $V(\omega + n)$ are the lists we are going to call SUPER- DUPER- NUMBERS:

$$V(\omega), PV(\omega), PPV(\omega), \ldots.$$

Super-duper-numbers are what tireless immortal beings encounter when they start with $V(\omega)$ (the list of all hereditarily finite lists) and apply the operation P to everything they encounter.

Exercise 4.8 *Using* Definition 3.9 *as your guide, say what it means for something to be a super-duper-number. Prove "super-duper" versions of Exercises 3.35 and 3.36.*

Exercise 4.9 *Show that every higher rank lists the entries of its entries.*

Since $\omega + Sn = S(\omega + n)$, Exercises 3.7 and 4.9 imply that the entries of $\omega + Sn$ are $\omega + n$ and the entries of $\omega + n$.

Exercise 4.10 *Show that every super-duper-number lists the entries of its entries. (You might consider a super-duper induction.)*

Exercise 4.11 *Show that if $n \neq 0$, then $\omega + n$ lists ω. (You might consider induction on n.)*

Exercise 4.12 *Show that n lists m only if $\omega + n$ lists $\omega + m$. (If you feel like another induction, you might consider doing it on n.)*

Exercise 4.13 *Show that no higher rank lists itself.*

Theorem 4.1 *If $\omega + n$ lists $\omega + m$, then n lists m.*

Proof Suppose $\omega + n$ lists $\omega + m$. Suppose m lists n. Then, as you showed in Exercise 4.12, $\omega + m$ lists $\omega + n$. So, according to Exercise 4.9, $\omega + m$ will list itself, contrary to Exercise 4.13. It will also contradict Exercise 4.13 if $m = n$. So Theorem 3.9 implies that n lists m.   □

We have confirmed that the listing relation orders the higher ranks the same way it does the numbers:

$$\omega + n \text{ lists } \omega + m \iff n \text{ lists } m.$$

So, in fact, the listing relation well-orders the higher ranks. (Recall Sect. 3.6) In particular, if some higher ranks satisfy a certain condition, there will be a unique *first* higher rank that does so.

Exercise 4.14 *Use induction (on n) to show that n lists m only if $V(\omega + n)$ lists $V(\omega + m)$.*

Exercise 4.15 *Show that no super-duper-number lists itself. (Theorem 3.17 might be useful.)*

Exercise 4.16 *Show that $V(\omega + n)$ lists $V(\omega + m)$ only if n lists m.*

We have confirmed that the listing relation orders the super-duper-numbers the same way it does the numbers:

$$V(\omega + n) \text{ lists } V(\omega + m) \iff n \text{ lists } m.$$

So, in fact, the listing relation well-orders the super-duper-numbers.

Our RANKS are the numbers and the higher ranks.

Exercise 4.17 *Show that if α and β are ranks, then $V(\beta)$ lists $V(\alpha)$ if and only if β lists α.*

Exercise 4.18 *Show that the listing relation well-orders our ranks.*

Definition 4.1 If x is a part of $V(\alpha)$ for some rank α, then $\rho(x)$ is the first such α.

Exercise 4.19 *What is $\rho(\omega)$?*

Definition 4.2 A list is ZERMELIAN if and only if it is an entry of a super-duper-number.[2]

We retain Definition 3.14 as our explanation of "outranking."

Theorem 4.2 *One Zermelian list outranks another if and only if some Zermelian entry of the former outranks each Zermelian entry of the latter.*

Proof Zermelian sets have all the properties of hereditarily finite sets that we used in our proof of Theorem 3.19. (It might be a good exercise to double-check that.) □

Definition 4.3 Some Zermelian lists are, collectively, of BOUNDED RANK if and only if a Zermelian list outranks all of them.

Exercise 4.20 *Show that any Zermelian lists of bounded rank will form a Zermelian list.*

Exercise 4.21 *Confirm that there is a Zermelian list with no Zermelian entries.*

Exercise 4.22 *Confirm that Zermelian lists with the same Zermelian entries are the same.*

Exercise 4.23 *Confirm that our ranks are Zermelian and list only ranks.*

Exercise 4.24 *Show that no Zermelian list lists every rank. (You might want to look at the proof of Theorem 3.17.)*

It may not be obvious why Theorem 4.2 and Exercises 4.20–4.24 are of any particular interest. Be patient. They will play a key role in our exploration of set theory in Chap. 5. Before turning to sets, however, we will use our profusion of lists to develop the foundations of calculus.

---

[2] The entries of the super-duper-numbers form a model of Zermelo's set theory Z. For a helpful discussion see Uzquiano [1] (http://www.jstor.org/stable/421182).

4.3 Real Numbers

We have been a little sloppy, saying "numbers" when we mean the natural numbers 0, 1, 2, 3, ... We are going to continue this practice even though we know perfectly well that mathematicians recognize other numbers. It is only our *dialect* that is quirky, however. Our *theory* does not confine us to the naturals. Indeed, our theory lets us show that some parts of ω, when suitably ordered, behave just like the (non-negative) real numbers. Before I can say which parts, I need to introduce some new terminology. Parts of ω that omit only finitely many numbers are said to be cofinite. A part of ω that omits only finitely many numbers will have a largest non-entry if it has any non-entries at all.

Definition 4.4 A part A of ω is COFINITE if and only if $\omega \backslash A$ has a largest entry or has no entries. (Recall that $\omega \backslash A$ is the list of numbers not listed by A.)

ω itself is cofinite because it omits no numbers. $\emptyset$ is not cofinite because it omits every number. The list of all numbers of the form $6a + 9b + 20c$ is cofinite because it lists every number larger than 43. The list of all numbers of the form $6a + 9b$ is not cofinite because it omits 3 and every number that is not a multiple of 3. We are going to focus on lists of this latter sort: the non-cofinite ones. If A is one of these lists, then $\omega \backslash A$ has entries, but no largest one.

Exercise 4.25 *Show that each entry of ω (each number) is a non-cofinite part of ω.*

Definition 4.5 An entry DISCRIMINATES between two lists if and only if it appears on one but not the other. Two lists AGREE on an entry if and only if it does not discriminate between them (that is, if and only if it appears on both or neither).

Definition 4.6 If A and B are non-cofinite parts of ω, we say $A < B$ (A PRECEDES B) if and only if $A \neq B$ and B lists the smallest number that discriminates between A and B.[3]

Note that we are placing the non-cofinite parts of ω in a certain *order*, but we are not comparing their *sizes*. For example, $\{1, 3\}$ has more entries than $\{0\}$. Yet $\{1, 3\} < \{0\}$, since $\{0\}$ lists the smallest number that discriminates between these two lists. Indeed, the list of all odd numbers precedes $\{0\}$ even though the former has infinitely more entries than the latter.

Exercise 4.26 *Show that one non-cofinite part of ω precedes all the others*

Exercise 4.27 *Show that if A is a non-cofinite part of ω, then $A < n$ for some number n.*

---

[3] Here is another way of expressing the same idea. If A is a part of ω and j is an entry of ω, let $A \cap j$ be the list whose entries are, exactly, the entries of A smaller than (i.e., listed by) j. Now suppose A and B are non-cofinite parts of ω. Then $A < B$ if and only if, for some number j, $B \backslash A$ lists j and $(A \cap j) = (B \cap j)$.

Exercise 4.28 *Show that if m and n are non-zero numbers, then* $\{m\} < n$.

Exercise 4.29 *Show that if m and n are numbers, then* $m < n$ *if and only if m is smaller than n (that is, n lists m).*

Exercise 4.30 *Suppose k is a number and A is a non-cofinite list of numbers. Show that if* $A < \{k\}$, *then* $A < \{k+1, k+2, \ldots, k+n\}$ *for some number n.*

My statement of the last exercise is not really kosher because I have not said what '+' means when applied to our numbers. Here is a more correct version: if $A < \{k\}$, then $A < j \backslash Sk$ for some number j. Recall that $Sk = \{0, 1, 2, \ldots, k\}$. So $j \backslash Sk$ is the result of lopping off all the numbers $0, 1, 2, \ldots, k$ from the front of j. That is, $j \backslash Sk$ has the form $\{k+1, k+2, \ldots, k+n\}$ as in Exercise 4.30. Note that we obtain

$$A < j \backslash Sk < \{k\}$$

no matter how close A is to $\{k\}$. So we can make $j \backslash Sk$ arbitrarily close to $\{k\}$ by picking a sufficiently large j. Or, to put it differently, if a list B squeezes between $j \backslash Sk$ and $\{k\}$

$$A < j \backslash Sk < B < \{k\}$$

then Exercise 4.30 allows us to leapfrog it by picking a big enough j':

$$A < j \backslash Sk < B < j' \backslash Sk < \{k\}.$$

We now turn to some particularly important facts about the relation $<$. A, B, and C are understood to be non-cofinite parts of ω.

Theorem 4.3 IRREFLEXIVITY: $A \not< A$.

Proof There is no way that $A \neq A$. □

Theorem 4.4 TRANSITIVITY: *If* $A < B < C$, *then* $A < C$.

Proof Let j be the smallest number that discriminates between A and B. Let k be the smallest number that discriminates between B and C. Then A and B agree on every number smaller than j, while B and C agree on every number smaller than k. B lists j, but A does not. C lists k, but B does not. Note that $j \neq k$. This leaves two cases. Case 1: j is smaller than k. Then, since B lists j, so does C. Since B and C agree on every number smaller than j, so do A and C. That is, j is the smallest number that discriminates between A and C. Case 2: k is smaller than j.[4] Then, since B does not list k, neither does A. Since A and B agree on every number smaller than k, so do A and C. That is, k is the smallest number that discriminates between A and C. □

Exercise 4.31 *Show that* $<$ *is* CONNECTED. *That is, if* $A \neq B$, *then either* $A < B$ *or* $B < A$.

---

[4] Yes, this is possible. Note that $\emptyset < \{2\} < \{1, 2\}$.

Theorem 4.5 DENSENESS: *If $A < C$, then $A < B < C$ for some finite B.*

Proof Let j be the smallest number that discriminates between A and C. Since A is not cofinite, we can let k be the smallest number larger than j that A does not list. Let B be the result of adding k to A while deleting all entries larger than k. Then B lists the smallest number that discriminates between A and B (namely, k) and C lists the smallest number that discriminates between B and C (namely, j). $\square$

I need to prove one more fact about $<$. First, though, some preliminaries. We let $[\emptyset, \omega)$ be the list of all non-cofinite parts of ω. The idea is that these parts form a ray with first entry $\emptyset$, but no last entry. We let $[\emptyset, A)$ be the list of non-cofinite parts of ω that precede A. The interval $[\emptyset, A]$ will be the result of adding A to $[\emptyset, A)$.

Definition 4.7 Suppose M is part of $[\emptyset, \omega)$. A is an UPPER BOUND of M if and only if M is part of $[\emptyset, A]$. If, in addition, no upper bound of M precedes A, then A is M's LEAST UPPER BOUND.

Now for our last fact about our ordering of $[\emptyset, \omega)$.

Theorem 4.6 COMPLETENESS: *Every part of $[\emptyset, \omega)$ with an upper bound has a least upper bound.*

Proof I will omit some details. Suppose M is a part of $[\emptyset, \omega)$ with an upper bound. Since, by Exercise 4.27, each such upper bound precedes a number, some number is an upper bound of M. Suppose 1 is the first such number. (You can check later to see whether our reasoning about this case applies more broadly.) If M lists 1, then none of M's upper bounds can precede 1 and, hence, 1 is M's least upper bound. Suppose M does not list 1. Then all of M's entries precede 1. We can use induction and some other tricks to show that if n is any number, then ω has parts $A_0, A_1, \ldots, A_n$ with the following properties. First:

$$A_0 = \emptyset.$$

Second:

$$A_{k+1} = \begin{cases} A_k & \text{if every entry of } M \text{ precedes } A_k \cup \{k\} \\ A_k \cup \{k\} & \text{otherwise.} \end{cases}$$

Let A be the part of ω whose entries are the numbers that appear on one of the lists A_n. In more conventional terminology:

$$A = \bigcup_{n \in \omega} A_n.$$

We are trying to creep up on M's least upper bound from below. We exclude from our big list A any number that would take us beyond all of M's entries. For example, we have the opportunity to add 0 at stage 1 of our construction, but decline to do so. Instead, $A_1 = A_0 = \emptyset$ since

$$(A_0 \cup \{0\}) = (\emptyset \cup \{0\}) = \{0\} = 1$$

and all of M's entries precede 1. Suppose, on the other hand, that some of M's entries do not precede $\{1\}$. Then adding 1 to our big list A would not take us too far. So we would say

$$A_2 = (A_1 \cup \{1\}) = (\emptyset \cup \{1\}) = \{1\}$$

and, hence, A would list 1. What happens as we move down the list? Could A list one number after another with no further omissions? First note: if none of M's upper bounds precede 1, then 1 is M's least upper bound and we are done. Suppose, then, that B is an upper bound that precedes 1. That is, $B < \{0\} = 1$. By Exercise 4.30, $B < \{1, 2, \ldots, n\}$ for some number n. All of M's entries would precede $\{1, 2, \ldots, n\}$. So, even if A listed every number from 1 to $n - 1$, it would omit n. We conclude that A omits a number greater than 1. Let k be the first such number. Then,

$$\{1, 2, \ldots, k - 1\} = A_k = A_{k+1}.$$

$\{1, 2, \ldots, k\}$ is an upper bound of M since all of M's entries precede $A_k \cup \{k\}$. If none of M's upper bounds precede $\{1, 2, \ldots, k\}$, we are done. On the other hand, if an upper bound does precede $\{1, 2, \ldots, k\}$, we can repeat the above argument to show that A omits a number greater than k. (Note, in particular, that

$$\{1, 2, \ldots, k - 1\} < B < \{1, 2, \ldots, k\}$$

only if

$$B < \{1, 2, \ldots, k - 1, k + 1, \ldots, k + j\}$$

for some number j.) This holds quite generally: after each omission there will be another omission. So A omits numbers, but there is no largest number it omits: A is not cofinite. Is A an upper bound of M? Suppose $A < C$ where C is an entry of M. Let i be the first number listed by C but not by A. Then $(A_j \cup \{j\}) < C$ whenever $i < j$ since C will list i but $A_j \cup \{j\}$ will not. So

$$A_{j+1} = (A_j \cup \{j\})$$

whenever $i < j$. Since this would make A cofinite, we conclude that each entry of M either precedes A or *is* A. That is, A is an upper bound of M. Suppose B is an upper bound of M that precedes A. Then A lists the smallest number that discriminates between A and B. Suppose this number is m. Then not all of M's entries precede $A_m \cup \{m\}$, but B *does* precede $A_m \cup \{m\}$. If $A_m \cup \{m\}$ is itself an entry of M, then B is not an upper bound of M. If $A_m \cup \{m\}$ precedes an entry of M, then B is not an upper bound of M. So B is not an upper bound of M after all. We conclude that, in fact, no upper bound of M precedes A. □

Exercise 4.31 and Theorems 4.3–4.6 state the characteristic properties of a linear continuum. We have arranged the non-cofinite parts of ω to look like the real numbers in the ray $[0, \infty)$. Since they form the same structure as the non-negative reals, it should not lead to any great confusion if we call them "real numbers." Under the Mirimanoff construction, each natural number is a non-cofinite part of ω (as you showed in Exercise 4.25). So all our natural numbers are real numbers. Which real numbers are they? Actually, a better question would be: which real numbers would we like them to be? We have infinitely many alternatives. One alternative is particularly attractive: we can let each natural number be itself. That is, we can let each natural number n be the real number n. So, for example, $\{0, 1, 2, 3\}$ can be the real number 4, while $\{0, 1, 2, 3, 4\}$ can be the real number 5.

If a non-cofinite part of ω happens to be a Mirimanoff number, we can now say what real number it is. (Itself!) What about all the other non-cofinite parts? Take $\{0, 1, 2, 3, 4, 7, 10\}$, for example. To determine the integer part of this real number, count how many consecutive numbers are listed starting from 0. That would be 5 (0 through 4). We now subtract the remaining numbers, one by one, from 5, raise 2 to each of those powers, and add the results to 5.

$$5 + 2^{5-7} + 2^{5-10} = 5 + 2^{-2} + 2^{-5} = 5.28125.$$

So $\{0, 1, 2, 3, 4, 7, 10\}$ is the real number 5.28125. What about $\{1, 2, 3, 4, 7, 10\}$? The technique we just described yields the following.

$$0 + 2^{0-1} + 2^{0-2} + 2^{0-3} + 2^{0-4} + 2^{0-7} + 2^{0-10} = 0.946289.$$

The integer part is 0 because, in $\{1, 2, 3, 4, 7, 10\}$, there are 0 consecutive numbers starting from 0. Note that Exercise 4.30 has the following special case. If $A < 1$, then

$$A < \sum_{i=1}^{n} \frac{1}{2^i} = \frac{1}{2} + \frac{1}{4} + \frac{1}{8} + \cdots + \frac{1}{2^n}$$

for some number n.

The proof of Theorem 4.6 may make a bit more sense now. Suppose M lists all the rational numbers whose square is less than 2. Then our construction of the lists A_n begins as follows:

$$A_1 = \quad \{0\}$$
$$A_2 = \quad \{0\}$$
$$A_3 = \quad \{0\}$$
$$A_4 = \quad \{0, 3\}$$
$$A_5 = \{0, 3, 4\}.$$

We add 0, 3, and 4 to our list because they do not take us beyond the square root of 2:

$$\begin{aligned}
\{0\} &= & 1 &< \sqrt{2} \\
\{0, 3\} &= & 1 + \tfrac{1}{4} &< \sqrt{2} \\
\{0, 3, 4\} &= & 1 + \tfrac{1}{4} + \tfrac{1}{8} &< \sqrt{2}.
\end{aligned}$$

We omit 1 and 2 because they do take us beyond the square root of 2:

$$\begin{aligned}
\{0, 1\} &= & 2 &> \sqrt{2} \\
\{0, 2\} &= & 1 + \tfrac{1}{2} &> \sqrt{2}.
\end{aligned}$$

So A_5 is the real number

$$1 + \frac{1}{4} + \frac{1}{8}$$

whose square is 1.890625. The idea is to add any power of $\frac{1}{2}$ that does not give our sum a square greater than 2. Skipping ahead:

$$A_9 = \{0, 3, 4, 6, 8\}.$$

So A_9 is the real number

$$1 + \frac{1}{4} + \frac{1}{8} + \frac{1}{32} + \frac{1}{128}$$

whose square is about 1.99957. The limiting value A is the smallest real number that surpasses all the entries of M, that is, all the rationals less than $\sqrt{2}$. More briefly, $A = \sqrt{2}$.

Exercise 4.32 *What real number is* $\{0, 1, 2, 6, 9, 14, 15, 16, 17, 18, 19\}$?

Exercise 4.33 *Show that the list of all odd numbers* $\{1, 3, 5, 7, \ldots\}$ *is the real number* $\frac{2}{3}$. *(If you get stuck, look up some facts about geometric series.)*

As I noted a couple pages back, $[\emptyset, \omega)$ (ordered by $<$) looks like $[0, \infty)$ (ordered by the usual "less than" relation). Why is this of any interest? Well, first, it tells us something about quantity: ω has at least as many parts as there are non-negative reals. (It has, in fact, exactly as many parts; but showing that requires a bit more argument.) It also tells us something about logical possibility: if our theory of lists is consistent, then our concept of a linear continuum is not fundamentally incoherent. On a more philosophical note, we have also taken a step toward establishing the unity of mathematics. Since theorems about the real numbers can be interpreted as theorems about lists, mathematicians who study the reals will, whether they know it or not, be helping us see what follows from our theory of lists. The set theoretic reductive program of the last century showed that virtually all mathematicians are working out the consequences of a single comprehensive theory. So even if you find your colleague's work quite incomprehensible, you can be confident you are laborers in the same vineyard.

Exercise 4.34 *What would go haywire if we dropped the qualification "non-cofinite" from our definition of the real numbers?*

4.4 From Lists to Sets

The *Oxford English Dictionary* (*OED*) recognizes two substantive uses of the term "set." The first ("set" as in "sunset") is remote from the mathematical sense. The second is somewhat closer: a number of things or persons set or placed together. For those who are curious about what this sort of "setting" might involve, the *OED* helpfully provides a discussion of the verb "to set" occupying more than six thousand lines! None of the twelve types of "setting" recognized by the *OED* involve an operation that would yield the sets of mathematics. This supports the following bland observation: mathematicians have borrowed the everyday word "set" and are using it as a technical term.

Many terms borrowed in this way acquire meanings quite distant from everyday language and experience. This does not mean it has to be an ordeal to learn the technical vocabulary. Some people just "get it" right away, as if they had a deep-seated predisposition to learn (or invent) the stuff. Others do just fine with the help of some analogies or similes or other little nudges. Set theory is treacherous ground because we innocently expect our everyday understanding of sets to be a rich source of helpful nudges and signposts. It is not.

Luckily, we can find help elsewhere: everyday language and experience make it easy for us to talk sensibly about unranked list-types; and unranked list-types behave (or can easily be imagined to behave) very much like mathematical sets. Mathematical sets are not part of our everyday world. Some may even find unranked list-types a bit weird. Not to worry: paper lists are certainly familiar. Our experience with paper lists makes it easy for us to gab about unranked list types. At that point, set theory itself is within walking distance. This is a blessing: it offers a smooth path into the heart of modern mathematics. The various forms of hoodwinking we professors have been practicing for years are, it turns out, entirely unnecessary.

4.5 Solutions of Odd-Numbered Exercises

4.1 By Exercise 3.25, entries of hereditarily finite lists are hereditarily finite. So $V(\omega)$ lists the entries of its entries and, hence, by Exercise 3.7, the precursors of $V(\omega)$ are $V(\omega)$ and the entries of $V(\omega)$. But $V(\omega)$ and its entries are all lists.

4.3 You might assume there is a list that lists $V(\omega)$ and, furthermore, lists Px whenever it lists x.

4.5 Each entry of $\emptyset$ is a number. Suppose n is a number whose entries are all numbers. By Theorem 3.7, the entries of Sn are n and the entries of n. So every entry of Sn is a number. We conclude that entries of numbers are numbers and, hence, ω lists the entries of its entries.

4.7 Say that a list x is a HIGHER RANK if and only if it satisfies the following condition: given any S-closed lists, if ω is one of them, then so is x. This immediately yields an induction principle: if some lists are S-closed and ω is one of them, then every higher rank is one of them. Our definition also guarantees that ω is a higher rank. Our first job is to show that if x is a higher rank, then $x = \omega + n$ for some number n. This is an easy induction because $\omega = \omega + 0$ and, if $x = \omega + n$, then $Sx = S(\omega + n) = \omega + Sn$. As for the converse, we need to show that $\omega + n$ is a higher rank whenever n is a number. $\omega + 0$ is a higher rank because ω is. Suppose $\omega + n$ is a higher rank x. Then $\omega + Sn = S(\omega + n) = Sx$. But our definition guarantees that Sx is a higher rank whenever x is.

4.9 In Exercise 4.5, you proved that ω lists the entries of its entries. Suppose x lists the entries of its entries. Then, according to Exercise 3.7, the precursors of x are x itself and x's entries. Suppose Sx lists y and y lists z. Then y is either x itself or an entry of x. In either case, z is a precursor of x and, so, is listed by Sx. That is, Sx lists the entries of its entries.

4.11 It is vacuously true that $\omega + 0$ has the desired property. Suppose $\omega + n$ lists ω. Since $\omega + n$ is an entry of $S(\omega + n)$ and $S(\omega + n) = \omega + Sn$, $\omega + n$ is an entry of $\omega + Sn$ and, hence, by Exercise 4.9, ω is an entry $\omega + Sn$.

4.13 If ω listed itself, it would be a number and, hence, a number would list itself—contrary to Exercise 3.16. Suppose x is a higher rank that does not list itself. Sx lists the precursors of x and, so, by Exercises 3.7 and 4.9, Sx lists x and the entries of x. Note that $Sx \neq x$ since x does not list x. If Sx were an entry of x, then x would be an entry of an entry of x—contrary to Exercise 4.9. So Sx is neither x nor an entry of x and, so, does not list itself.

4.15 By Exercise 3.27, $V(\omega)$ lists every number. If $V(\omega)$ listed itself, it would be hereditarily finite and, hence, a hereditarily finite list would list every number—contrary to Theorem 3.17. Suppose Px lists itself. Then Px is a part of x and, hence, x is an entry of x since x is an entry of Px. We conclude: if x does not list itself, then neither does Px.

4.17 If α and β are both numbers, we apply Exercises 3.37 and 3.38. If α and β are both higher ranks, we apply Exercises 4.14 and 4.16. Suppose α is a number and β is a higher rank. By Exercises 4.9 and 4.11, β lists α. By Exercises 3.19 and 4.14, $V(\beta)$ either lists $V(\omega)$ or is $V(\omega)$. $V(\alpha)$ is hereditarily finite since it is an entry of $V(S\alpha)$. So $V(\omega)$ lists $V(\alpha)$ and, hence, by Exercise 4.10, $V(\beta)$ lists $V(\alpha)$.

4.19 By Exercise 3.27, ω is a part of $V(\omega)$. If ω were a part of $V(n)$, n a number, then ω would be an entry of $V(n + 1)$—contrary to Theorem 3.17. So ω does not list any rank α such that ω is a part of $V(\alpha)$ and, hence, $\rho(\omega) = \omega$.

4.21 $\emptyset$ is hereditarily finite because it is an entry of the super-number $P\emptyset$. So $\emptyset$ is an entry of the super–duper–number $V(\omega)$ and, hence, is Zermelian. Since $\emptyset$ has no entries, it has no Zermelian entries.

4.23 By Exercise 3.27, each number is an entry of $V(\omega)$ and, hence, is Zermelian. We are going to use induction to confirm that each higher rank $\omega + n$ is an entry of $V(\omega + Sn)$ and, hence, is Zermelian. Since $\rho(\omega) = \omega$ (Exercise 4.19), ω is an entry of $V(\omega + S0)$. As an inductive hypothesis, suppose $V(\omega + Sn)$ lists $\omega + n$. By Exercises 3.7, 4.9, and 4.10, every precursor of $\omega + n$ is an entry of $V(\omega + Sn)$. So $S(\omega + n)$ is a part of $V(\omega + Sn)$ and, hence, is an entry of $PV(\omega + Sn)$. But $S(\omega + n) = (\omega + Sn)$ and $PV(\omega + Sn) = V(\omega + SSn)$. So $\omega + Sn$ is an entry of $V(\omega + SSn)$. Now we need to confirm that every rank is a list of ranks. $\emptyset$ only lists ranks because it lists nothing. ω only lists ranks because it only lists numbers. Suppose rank α only lists ranks. By Theorem 3.6 and Exercises 3.7 and 4.9, $S\alpha$ only lists α and the entries of α. So $S\alpha$ only lists ranks.

4.25 $\emptyset$ is a part of ω because it is a part of every list. $\omega \backslash \emptyset$ is ω. ω has no largest entry because each number n is an entry of Sn. So $\emptyset$ is a non-cofinite part of ω. As an inductive hypothesis, suppose number n is a non-cofinite part of ω. By Theorem 3.6 and Exercise 3.7, Sn is a part of ω. Suppose m is the largest entry of $\omega \backslash Sn$. Then Sn lists Sm but does not list m—contrary to Theorem 3.6. We conclude that Sn is non-cofinite.

4.27 Since A is non-cofinite, $\omega \backslash A$ has entries. By Theorem 3.8, we can let k be the smallest of those entries. k is the smallest number not listed by A. k is listed by Sk. So Sk lists the smallest number that discriminates between A and Sk. That is, $A < Sk$.

4.29 By Exercise 3.16, m does not list m. m lists all smaller numbers (since "smaller than" means "listed by"). So m is the first number not listed by m. If n lists m, then n lists the smallest number that discriminates between m and n. On the other hand, suppose n lists k but m does not. Suppose $k \neq m$. By Theorem 3.9, k lists m and, hence, by Theorem 3.6, n lists m.

4.31 Suppose $A \neq B$. By Axiom 3.1, A and B do not have the same entries and, hence, there are numbers that discriminate between them. By Theorem 3.8, there is a smallest such number. The list that does not list that number will precede the one that does.

4.33

$$\sum_{i=0}^{\infty} \left(\frac{1}{2}\right)^{2i+1} = \frac{1}{2} \sum_{i=0}^{\infty} \left(\frac{1}{2}\right)^{2i} = \frac{1}{2} \sum_{i=0}^{\infty} \left(\frac{1}{4}\right)^{i} = \frac{1}{2} \left(\frac{1}{1 - \frac{1}{4}}\right) = \frac{1}{2} \cdot \frac{4}{3} = \frac{2}{3}.$$

References

1. Uzquiano, G. (1999). Models of second-order Zermelo set theory. *Bulletin of Symbolic Logic*, *5*, 289–302.
2. Zermelo, E. (1930). Über Grenzzahlen und Mengenbereiche. *Fundamenta Mathematicae*, *16*, 29–47.

Chapter 5
The Hierarchy of Sets

5.1 New Axioms

We will now rejoin the rest of the mathematical community and say "set" rather than "list", "member" rather than "entry", "empty set" rather than "blank list", "subset" rather than "part", and so on. We are also going to start afresh with new axioms characterizing ranks and their relationship to sets. We will not, however, forsake our friends the lists. We will use our old list theory to show that our new set axioms can be interpreted as claims about lists. We will also use our new set theory to show that our old list axioms can be interpreted as claims about sets. This may seem an odd way to spend our time, first going to the trouble of introducing new material and then going to even more trouble to show that this material is not so new after all. What a letdown to roll out shiny new vocabulary and axioms and then reveal that the new stuff is just an offbeat way of talking about stale old stuff!

Well, the situation is hardly as bad as all that. First, it is a commonplace that you can get fresh insights by approaching something familiar from a new direction. But, second, there is a more specific reason for our preoccupation with interpretability. Sorting out relations of interpretability (what is interpretable as what) is a mathematical enterprise that helps us understand how various bits and pieces fit together to form the great edifice of mathematics. Interpretation is a mathematical tool for exploring the architecture of mathematics. For example, if we show that all the settled results of mathematical field F are interpretable as assertions derivable from theory T, we can conclude that T provides a foundation for F. Among other things, this would mean that all the ideas and assumptions deployed in F can be reconstructed within T (just as, in the previous chapter, we reconstructed the principles of real number analysis inside our list theory). So, even if an evil mathematics czar outlawed everything that bears the outward stamp of F, the mathematically essential insights and methods of F might be preserved in T. If we showed that all settled results in *every* mathematical sub-field can be interpreted as assertions derivable from T, we could conclude that T provides a foundation for all of mathematics. Philosophers who want to understand mathematics need to understand these architectural analyses: both how

S. Pollard, *A Mathematical Prelude to the Philosophy of Mathematics*,
DOI: 10.1007/978-3-319-05816-0_5,
© Springer International Publishing Switzerland 2014

they are performed and what they reveal. So another round of interpretability results seems warranted. First, though, we need our new axioms. Here are some definitions to prepare the way.

Definition 5.1 One rank PRECEDES another if and only if the first is a member of the second.

In symbols, using the Greek letter epsilon ('$\in$') to express the membership relation and, as in the last two chapters, letting the Greek letter rho ('ρ') stand for the rank function:

$$\rho(x) \text{ precedes } \rho(y) \text{ if and only if } \rho(x) \in \rho(y).$$

When we discuss the ordering of our ranks, we are talking about which ranks are members of which. This should not seem so strange after your experience with list-theoretic numbers that were ordered by the listing relation. When we discussed the ordering of our numbers, we were talking about which numbers were entries of which.

Definition 5.2 A set y OUTRANKS a set x if and only if $\rho(x)$ precedes $\rho(y)$.

Theorem 3.19 is the inspiration for our first new axiom. We are going to drop the business about hereditary finiteness and translate the list talk into set talk. Furthermore, we are going to turn our reasoning process upside down, treating our former theorem as a fundamental premise, an axiom, and deriving facts about sets from it.[1] It will simplify our task if we describe a universe inhabited only by sets. Within the confines of our theory, claims about "every set" will be assertions about everything in our limited universe.

Axiom 5.1 One set outranks another if and only if some member of the former outranks each member of the latter.

We will adopt more axioms after some preliminary definitions.

Definition 5.3 Some sets FORM a set if and only if the former sets are exactly the members of the latter set.

If sets x and y were to form a set, it would be $\{x, y\}$, the set whose members are x and y.

Definition 5.4 Some sets are, collectively, of BOUNDED RANK if and only if there is a set that outranks all of them.

Axiom 5.2 Any sets of bounded rank will form a set.

Axiom 5.3 There is a set with no members.

---

[1] This clever idea comes from Van Aken [9] (http://www.jstor.org/stable/2273911). For a discussion of our axioms, see Pollard [6], pp. 161–164, and [7].

How many sets have no members? The next axiom tells us there is exactly one.

Axiom 5.4 Sets with the same members are the same.

We can now let $\emptyset$ be the one and only memberless set. Axioms 5.3 and 5.4 do not say anything new or amazing. They are just Axioms 3.1 and 3.2 translated into the language of sets.

Exercise 5.1 *Show that $\emptyset$ outranks no set. Show that every set other than $\emptyset$ outranks $\emptyset$.*

After yet another definition, we will use our axioms to prove a fundamental theorem.

Definition 5.5 Some sets, the X's, are collectively CLOSED UNDER SET FORMATION if and only if any set formed by some of the X's is itself one of the X's.

If there is such a set as $\{x, y\}$ and this set is distinct from both x and y, then x and y are not closed under set formation: x and y form $\{x, y\}$, but $\{x, y\}$ is neither x nor y. You have to go beyond x and y to encounter the set formed by x and y.

Theorem 5.1 *If some sets are closed under set formation and $\emptyset$ is one of them, then every set is one of them.*

Proof Suppose some sets are closed under set formation. Let's call them "the V-sets" or "the V's". All the other sets will be "the non-V's". Suppose $\emptyset$ is one of the V's, while x is one of the non-V's. Note that every non-V outranks $\emptyset$. Furthermore, $\emptyset$ is not a member of itself (since $\emptyset$ has no members at all). So there is at least one set that (1) is outranked by every non-V and (2) is not a member of itself. By Axiom 5.2, all the sets satisfying these conditions will form a set. Let y be this set. Now consider x: our representative of the non-V's. If all of x's members were V-sets, then x would be a V-set too (since the V-sets are closed under set formation). So we can let w be a member of x that is a non-V.[2] Since w outranks every member of y, Axiom 5.1 guarantees that x outranks y. The only information we used about x is that it is a non-V. So we can conclude that every non-V outranks y. If y is not a member of itself, then it satisfies both conditions for membership in y and, hence, *is* a member of itself. But if y is a member of itself, it must satisfy the second condition for membership in y and, hence, is not a member of itself. So y is a member of itself if and only if it isn't. Since this is absurd, we must have gone wrong somewhere. Our misstep was back at the start when we assumed that there are non-V's. $\square$

The preceding theorem is known as the principle of EPSILON- INDUCTION. If you can show that $\emptyset$ has an infection and that every set with infected members is infected, then you can show that every set is infected.

---

[2] What if there were non-sets in our universe? Then we would have no guarantee that w is a set. Since non-V's are *sets* that are not V's, we would have no reason to believe that w is a non-V. On the other hand, if we allowed non-sets to be non-V's, we could not have inferred that every non-V outranks $\emptyset$.

Exercise 5.2 *Use epsilon-induction to show that every set outranks its members.*

Exercise 5.3 *Use epsilon-induction to show that no set outranks itself.*

$x \in x$ only if $\rho(x) \in \rho(x)$. So no set is a member of itself.

Theorem 5.2 *If x outranks y and y outranks z, then x outranks z.*

Proof Exercise 5.1 confirms that the theorem is true when $z = \emptyset$. (If x outranks y, then $x \neq \emptyset$ and, hence, x outranks $\emptyset$.) As an inductive hypothesis, suppose the theorem comes out true whenever we let z be a member of set s. We want to show that the theorem is true when $z = s$. Suppose, then, that x outranks y and y outranks s. By Axiom 5.1, we can let x' be a member of x that outranks every member of y, while y' is a member of y that outranks every member of s. Note that x' outranks y'. So, by our inductive hypothesis, x' outranks every member of s. Hence, by Axiom 5.1, x outranks s. Since $\emptyset$ satisfies the theorem and the sets satisfying the theorem are, collectively, closed under set formation, we conclude that every set satisfies the theorem.   □

Definition 5.6 α is a RANK if and only if $\alpha = \rho(x)$ for some set x.

In what follows, we will assume that α, β, γ are ranks.

Theorem 5.3 IRREFLEXIVITY: $\alpha \notin \alpha$.

Theorem 5.4 TRANSITIVITY: *If $\alpha \in \beta$ and $\beta \in \gamma$, then $\alpha \in \gamma$.*

Theorem 5.5 WELL-FOUNDEDNESS: *Given any ranks at all, one will be preceded by none of the others.*

Proof Pick some ranks and call them "the X-ranks". All other sets will be "non-X's". If $\emptyset$ is an X-rank, then an X-rank is preceded by no X-rank, as desired. Suppose $\emptyset$ is a non-X. If, in addition, a set is a non-X whenever its members are non-X's, then every set is a non-X, contrary to our assumption that there are X-ranks. So there must be an X-rank whose members are all non-X's. But this means there is an X-rank preceded by no X-ranks.   □

Theorem 5.6 *If x outranks every member of y, then y does not outrank x.*

Proof Suppose x is a counter-example to the theorem. This means we can pick a y that satisfies the following condition:

x outranks every member of y even though y outranks x.

By Axiom 5.1, some member of y outranks every member of x. This means we can pick a member z of y that satisfies the following condition:

z outranks every member of x even though x outranks z.

Such a z would be a counter-example to the theorem outranked by our original counter-example x. Since this means no counter-example has minimal rank, it would contradict Theorem 5.5 for there to be any counter-example at all.   □

Recall that, in the Mirimanoff construction Ø, SØ, SSØ, ..., each number is the list of prior numbers. We now assume that ranks have a similar property.

Axiom 5.5 Every rank is a set of ranks.

Exercise 5.4 *Show that Ø is a rank. (You might consider whether $\rho(\emptyset)$ has any members.)*

Theorem 5.7 CONNECTEDNESS: *Given any two ranks, one will precede the other.*

Proof Say that a rank is DISCONNECTED if some other rank neither precedes nor is preceded by it. Suppose there is a disconnected rank. Then we can use Theorem 5.5 to pick a disconnected rank α preceded by no disconnected rank. We say α is a MINIMAL disconnected rank. Since α is disconnected, there is a rank distinct from α that neither precedes nor is preceded by α. Use Theorem 5.5 to let γ be a minimal such rank. If a rank β precedes γ, β will either precede or be preceded by α. (β cannot *be* α because α would then precede γ.) If β is preceded by α, then, by Theorem 5.4, α will precede γ. ($\alpha \in \beta \in \gamma \Rightarrow \alpha \in \gamma$.) So β will precede α and, more generally, every rank that is a member of γ will be a member of α. According to Axiom 5.5, this means that every member of γ is a member of α. Similar reasoning shows that every member of α is a member of γ. So, by Axiom 5.4, $\alpha = \gamma$, contrary to our assumption that they are distinct. We conclude that there is no disconnected rank.□

Theorem 5.8 $\rho(\alpha) = \alpha$.

Proof Suppose α is the first rank that violates the theorem. Let $\alpha = \rho(x)$. Then:

$$\beta \in \alpha \implies \rho(\beta) = \beta \implies \rho(\beta) \in \alpha \implies \rho(\beta) \in \rho(x).$$

So x outranks every member of α and, hence, by Theorem 5.6, α does not outrank x. That is, $\rho(x) \notin \rho(\alpha)$ and, hence, $\alpha \notin \rho(\alpha)$. Suppose $\rho(\alpha) \in \alpha$. Then, by the minimality of α, $\rho(\rho(\alpha)) = \rho(\alpha)$. Furthermore, by Exercise 5.2, $\rho(\rho(\alpha)) \in \rho(\alpha)$. So $\rho(\alpha) \in \rho(\alpha)$, contrary to Exercise 5.3. We conclude that $\rho(\alpha) \notin \alpha$. So, by Theorem 5.7, $\rho(\alpha) = \alpha$. This means no counter-example to the theorem can have minimal rank. So Theorem 5.5 guarantees there are no counter-examples at all. □

Exercise 5.5 *Show that $\rho(\rho(x)) = \rho(x)$.*

Exercise 5.6 *Show that β outranks α if and only if $\alpha \in \beta$.*

Definition 5.7 A set x is a SUBSET of a set y (in symbols, $x \subseteq y$) if and only if every member of x is a member of y.

Exercise 5.7 *Show that if $\alpha \neq \beta$ and $\alpha \subseteq \beta$, then $\alpha \in \beta$.*

Definition 5.8 A set x is TRANSITIVE if and only if every member of x is a subset of x (that is, $z \in y \in x$ only if $z \in x$).

Exercise 5.8 *Show that every rank is transitive.*

Axiom 5.5 and Exercise 5.8 guarantee that every rank is a transitive set of ranks. Now we want to prove the converse.

Exercise 5.9 *Show that if x is a set of ranks, then* $x \subseteq \rho(x)$.

Theorem 5.9 *Every transitive set of ranks is a rank.*

Proof Suppose x is a transitive set of ranks. We want to show that x has the same members as $\rho(x)$. Suppose $\alpha \in \rho(x)$. Then, by Theorem 5.8, $\rho(\alpha) \in \rho(x)$. That is, x outranks α. So, by Theorem 5.6, α cannot outrank every member of x. Suppose β is a member of x not outranked by α. Then, by Exercise 5.2, $\beta \notin \alpha$. So, by Theorem 5.7, either $\alpha \in \beta$ or $\alpha = \beta$. In the each case, $\alpha \in x$. (In the first case, because x is transitive.) We conclude that $\rho(x) \subseteq x$. Now apply Exercise 5.9 and Axiom 5.4. $\square$

5.2 Two Models

At the beginning of this chapter, I noted that we will explore interpretability relations *between* theories. We will consider interpretations or translations that carry us from one theory to another. We can treat this as an entirely syntactic exercise, assigning sentences to sentences without ever indicating what those sentences mean. I hope it will not be too confusing if we follow established usage and (as in Sect. 2.3) talk about "interpretations" of another sort, interpretations that offer readings of individual theories rather than mappings between theories. In such a reading of our set theory, we would (1) specify our UNIVERSE OF DISCOURSE (we would indicate what we are discussing when we say "every set", "any sets," "some set", or "some sets"); (2) indicate which objects in our universe are MEMBERS of which objects; and (3) indicate which objects in our universe are the RANKS of which objects. Recall, from Chap. 2, that a MODEL of our set theory would be an interpretation that makes all our axioms true. At the moment, our only axioms are 5.1–5.5 (though we will add more in a bit).

Here is a very simple interpretation. Let our universe of discourse consist of the lamp on my desk. When we say that *all* or *some* objects in the universe have a certain property, we will just mean that the lamp has that property. The lamp will count as a set, it will have no members, and its rank will be itself.

If we eliminate defined vocabulary and translate things into a kind of logicianese, Axiom 5.1 reads as follows.

> Every set x and every set y is such that $[\rho(x) \in \rho(y)$ if and only if some member z of y is such that [every member w of x is such that $[\rho(w) \in \rho(z)]]]$.

If that looks like gibberish, a more formal version may be helpful.

$$\forall x \forall y ((x \text{ is a set } \wedge \ y \text{ is a set}) \rightarrow (\rho(x) \in \rho(y) \leftrightarrow \exists z \in y \forall w \in x \ \rho(w) \in \rho(z))).$$

According to the above interpretation, this says that my lamp is a member of my lamp if and only if some member of my lamp has a certain property that we need not analyze further. Since my lamp has no members, both sides of this biconditional are false and, hence, the biconditional itself is true.[3] So our interpretation makes Axiom 5.1 true.

Exercise 5.10 *Show that our interpretation makes Axioms 5.2–5.5 true.*

You have confirmed that our interpretation is a model of our set theory: we can make all our axioms true in a universe that has only one thing in it. So our theory, though certainly consistent, is not yet of much mathematical use: it does not assert the existence of things that behave like the natural numbers or the real numbers or other interesting objects of mathematical investigation. In the next section, we shall consider how to strengthen our theory. First, however, we consider another interpretation of our theory.

Once again, we let my lamp be the only object in our universe and we let my lamp's rank be my lamp. This time, though, we give my lamp a member: itself. Let λ be my lamp. Then we have assumed the following: $\lambda \in \lambda$ and $\rho(\lambda) = \lambda$. Since $\rho(\lambda) \in \rho(\lambda)$, λ is of bounded rank. So Axiom 5.2 says our universe must feature a set $\{\lambda\}$ whose only member is λ. Indeed it does, because $\lambda = \{\lambda\}$.

Exercise 5.11 *Confirm that our new interpretation makes Axioms 5.1, 5.2, 5.4, and 5.5 true, but makes Axiom 5.3 false.*

You have just shown that Axiom 5.3 is INDEPENDENT: it does not follow from the other axioms of our theory. If it did follow, it would not be possible for it to be false while the other axioms are true. But you just showed that this *is* possible.

5.3 How Many Ranks?

Axiom 5.2 is only too happy to give us lots of sets as long as we hold up our end: we need to be good at showing that sets are of bounded rank. It will be easier for us to do that if there are lots of ranks. For example, if there is a non-empty rank, it will outrank $\emptyset$. Since $\emptyset$ will then be of bounded rank, there will be a set whose only member is $\emptyset$. That is, Axiom 5.2 will give us $\{\emptyset\}$. If there are two non-empty ranks, the larger of them will outrank both $\emptyset$ and $\{\emptyset\}$ and Axiom 5.2 will give us $\{\emptyset, \{\emptyset\}\}$ and $\{\{\emptyset\}\}$. Here is an axiom that gives us infinitely many ranks.

Axiom 5.6 Every rank precedes some rank.

I am not saying that some rank is preceded by every rank. (It would then have to precede itself.) The point is, on the contrary, that there is no maximum rank. Given any rank α, there is a rank preceded by α.

Exercise 5.12 *Show that each set is outranked by some set.*

---

[3] A biconditional is an "if and only if" statement. In classical logic, a biconditional $\ulcorner \phi \leftrightarrow \psi \urcorner$ is true when ϕ and ψ are both false (or both true).

Exercise 5.13 *Show that any sets* x, y *will form a set* $\{x, y\}$.

Exercise 5.14 *Show that any members of a set will form a set.*

Exercise 5.15 *Show that if* x *is a set, then the members of* x's *members form a set* $\bigcup x$

Exercise 5.16 *Show that the subsets of any set* x *form a set* Px.

You have just derived four of the axioms of Zermelo's set theory Z (mentioned in Chaps. 3 and 4). Our Axiom 5.4 is another Z axiom.

Theorem 5.10 *The members of any sets* x, y *form a set* $x \cup y$.

Proof Note that $x \cup y = \bigcup \{x, y\}$.   □

Definition 5.9 $\alpha + 1 = \alpha \cup \{\alpha\}$.

Exercise 5.17 *Show that* $\alpha + 1$ *is the first rank preceded by* α.

Note that the sets outranked by α are of bounded rank and, so, form a set.

Definition 5.10 $V(\emptyset) = \emptyset$. If α outranks any sets, then the members of $V(\alpha)$ are the sets outranked by α.

That is, $x \in V(\alpha)$ if and only if $\rho(x) \in \alpha$. So $x \in V(\rho(y))$ if and only if y outranks x.

Exercise 5.18 *Show that* $V(\alpha)$ *is transitive.*

Exercise 5.19 *Show that* $V(\alpha + 1) \subseteq PV(\alpha)$.

Exercise 5.20 *Show that* $PV(\alpha) \subseteq V(\alpha + 1)$. (*It might help if you suppose* $\alpha = \rho(z)$.)

Theorem 5.11 $V(\alpha + 1) = PV(\alpha)$.

Exercise 5.21 *Show that if* $\alpha \in \beta$, *then* $V(\alpha) \subseteq V(\beta)$.

Exercise 5.22 *Show that if* $\alpha \in \beta$, *then* $V(\alpha) \in V(\beta)$.

Theorem 5.12 $\rho(x)$ *is the first* α *such that* $x \subseteq V(\alpha)$.

Proof By Exercises 5.2 and 5.5, $y \in x$ only if $\rho(y) \in \rho(\rho(x))$. So $\rho(x)$ outranks every member of x and, hence, $x \subseteq V(\rho(x))$. Suppose $x \subseteq V(\alpha)$. Then α outranks every member of x. So, by Theorem 5.6, x does not outrank α and, hence, by Theorem 5.7, $\rho(x) \in \alpha$ or $\rho(x) = \alpha$, as desired.   □

Theorem 5.13 $\rho(x) + 1$ *is the first* α *such that* $x \in V(\alpha)$.

Proof Theorems 5.11 and 5.12 let us reason as follows:

$$x \subseteq V(\rho(x)) \Longrightarrow x \in PV(\rho(x)) \Longrightarrow x \in V(\rho(x) + 1).$$

According to Exercise 5.17, $\rho(x) + 1$ is the first rank preceded by $\rho(x)$. So $\rho(x) \in \alpha$ only if $\alpha \notin \rho(x) + 1$. But $x \in V(\alpha)$ only if $\rho(x) \in \alpha$. So $x \in V(\alpha)$ only if $\alpha \notin \rho(x) + 1$. That is, $\rho(x) + 1$ is minimal. $\square$

Exercise 5.18 and Theorem 5.13 yield the following.

Theorem 5.14 *Every set is a member of a transitive set.*

Exercise 5.23 *Show that $\rho(V(\alpha)) = \alpha$.*

Exercise 5.24 *Show that $V(\alpha) \in V(\beta)$ only if $\alpha \in \beta$.*

5.4 Equiconsistency

We turn now to the task of showing that our new set theory is interpretable in our old list theory—and vice versa. Here our interest is *inter-theoretic* interpretation. We will consider translations that carry us from one theory to another. Recall our definition of 'Frege-precursor' from Chap. 3: x is a Frege-precursor of y if and only if, given any entry-closed lists, if y is one of those lists, then so is x. We now translate this definition from the language of list theory into the language of set theory.

Definition 5.11 Some sets are, collectively, MEMBER- CLOSED if and only if each member of one of them is itself one of them.

Definition 5.12 x is a FREGE- PRECURSOR of y if and only if, given any member-closed sets, if y is one of those sets, then so is x.

We will now consider a new definition of 'precursor' and will show that the precursors of a set are exactly its Frege-precursors.

Definition 5.13 x is a PRECURSOR of y if and only if x is a member of every transitive set that has y as a member.

Suppose x is a precursor of y. We want to show that x is also a Frege-precursor of y. Pick some sets that are, collectively, member-closed. Call them "the M-sets". Then every member of an M-set is an M-set. Suppose y is an M-set. We want to show that x is an M-set too. Consider the M-sets that do not outrank y. Since these sets are of bounded rank, they form a set z.

Exercise 5.25 *Show that the set z described above is transitive.*

Since y does not outrank itself, it is a member of z. So, by Definition 5.13, x is a member of z and, hence, x is an M-set. We conclude that x will appear among some member-closed sets whenever y does. That is, x is a Frege-precursor of y. More generally, every precursor of a set is a Frege-precursor. We now want to prove the converse.

Suppose x is a Frege-precursor of y. Suppose y is a member of the transitive set z. Then z's members, collectively, have the property that every member of one of them is itself one of them. That is, z's member are, collectively, member-closed. So, by Definition 5.12, x is one of z's members. We conclude that x will be a member of a transitive set whenever y is. That is, x is a precursor of y.

Set precursors are just like list precursors: the precursors of a set y are what tireless immortal beings encounter when they start with y and persist in tracking down the members of everything they encounter. Precursors of y are y itself, the members of y, the members of y's members, and so on.

Exercise 5.26 *Show that the precursors of a set x form a set Sx . (You might start by confirming that each precursor of x is a member of $V(\rho(x) + 1)$).*

Exercise 5.27 *Show that if y is transitive, then $Sy = y \cup \{y\}$. (You might first show that $y \cup \{y\}$ is transitive.)*

Note that, in particular, $S\alpha = \alpha + 1$.

Theorem 5.15 *Any sets of precursors of a set will form a set.*

Proof Any set of precursors of a set x will be a subset of Sx and, hence, a member of PSx. Now apply Exercise 5.14.   □

When we translate Axioms 3.1–3.3 from the language of list theory into the language of set theory we obtain Axiom 5.3, Axiom 5.4, and Theorem 5.15. If ϕ is a sentence in the language of list theory that follows from Axioms 3.1–3.3, then the translation of ϕ into the language of set theory will follow from Axiom 5.3, Axiom 5.4, and Theorem 5.15 and, hence, will follow from Axioms 5.1–5.6 (since Theorem 5.15 follows from Axioms 5.1–5.6). So every result we obtained from Axioms 3.1–3.3 becomes, when suitably translated, a theorem of our set theory and we can treat it as such without bothering to give a whole new proof.

Suppose Axioms 3.1–3.3 are inconsistent. Then every sentence in the language of list theory follows from them. This would include the absurdity:

Some list both is and is not an entry of itself.

Our translation of this sentence into the language of set theory is:

Some set both is and is not a member of itself.

Since this sentence is absurd, we conclude that Axioms 3.1–3.3 are inconsistent only if Axioms 5.1–5.6 are. So any evidence for the consistency of Axioms 5.1–5.6 is evidence for the consistency of Axioms 3.1–3.3. We say Axioms 3.1–3.3 are consistent "relative to" Axioms 5.1–5.6. We also say that Axioms 3.1–3.3 are INTERPRETABLE IN Axioms 5.1–5.6.

We have just seen how to interpret the list-theoretic claims of Axioms 3.1–3.3 as set-theoretic claims that follow from Axioms 5.1–5.6. We now want to reverse course and show how to interpret the set-theoretic claims of Axioms 5.1–5.6 as list-theoretic claims that follow from Axioms 3.1–3.3. We read "set" as "hereditarily finite list" and "member of" as "entry of". We use Definition 3.13 to interpret the notion of rank. Theorem 3.19, then, is our translation of Axiom 5.1. Exercise 3.43 is our translation of Axiom 5.2. Axiom 5.3 presents no difficulties: the blank list is a hereditarily finite list with no hereditarily finite entries. Exercise 3.32 is our list-theoretic version of Axiom 5.4. Our list-theoretic interpretation of "rank" makes Axioms 5.5 and 5.6 true because every number is a list of numbers (Exercise 4.5) and every number is an entry of a number. So we have shown how to make Axioms 5.1–5.6 come out true in the universe of hereditarily finite lists. If ϕ is a sentence in the language of set theory that follows from Axioms 5.1–5.6, then the translation of ϕ into the language of list theory will follow from Axioms 3.1–3.3. So every result we obtained from Axioms 5.1–5.6 becomes, when suitably translated, a theorem of our list theory and we can treat it as such without bothering to give a whole new proof.

If it follows from Axioms 5.1–5.6 that every set both is and is not a member of itself, then it will follow from Axioms 3.1–3.3 that every hereditarily finite list both is and is not an entry of itself. Axioms 5.1–5.6 are inconsistent only if Axioms 3.1–3.3 are. So any evidence for the consistency of Axioms 3.1–3.3 is evidence for the consistency of Axioms 5.1–5.6. Having proved RELATIVE CONSISTENCY in both directions, we say that Axioms 3.1–3.3 and Axioms 5.1–5.6 are EQUICONSISTENT. We have shown: Axioms 3.1–3.3 are consistent if and only Axioms 5.1–5.6 are.

5.5 Even More Ranks

Our first interpretability result from the previous section allows us to exploit the number theoretic results of Chap. 3. In particular, we can assume that the numbers

$$\emptyset, S\emptyset, SS\emptyset, \ldots$$

that is,

$$\emptyset, \{\emptyset\}, \{\emptyset, \{\emptyset\}\}, \ldots$$

are well-defined and obey an induction principle. Given induction, Exercises 5.4, 5.17, and 5.27 imply that each number is a rank. What about the converse? Is every rank a number? Suppose some rank is not and use Theorem 5.5 to let β be the first such rank. This means β is a set of numbers. Since $\emptyset$ is a number, but β is not, β is non-empty. Exercises 5.4 and 5.7 imply that $\emptyset$ is a member of every non-empty rank. So $\emptyset \in \beta$. Suppose the number n is a member of β, but Sn is not. Since β is transitive, Sn cannot be a member of any member of β. Given any two numbers, one will be a member of the other. So every member of β is a member of Sn. On the other hand, the members of Sn are n and the members of n and these are all members of β. But this means $\beta = Sn$, contrary to our assumption that β is not a number. We conclude that Sn is a member of β whenever n is and, more generally, that every number is a member of β. A rank that is not a number will have every number as a member.

We are going to translate this result into the language of list theory so we can apply the second interpretability result from the previous section. To be sure we know what we are translating, let us express our result more formally. On the assumption that some rank x is not a number, we have found that x will, first of all, have the following property.

$$\exists y(y \in x \land \forall z\, z \notin y)$$

Our translation of this will say that x is a hereditarily finite list with a hereditarily finite entry y that has no hereditarily finite entries. Since super-numbers list the entries of their entries, entries of hereditarily finite lists are themselves hereditarily finite. So y will have no entries at all and our translation will say that x lists the blank list. Returning to the world of sets, the set x (the rank that is not a number) will also have the following property.

$$\forall w(w \in x \to w \cup \{w\} \in x)$$

In the language of set theory, the term '$w \cup \{w\}$' refers to the set whose members are w and the members of w. If w were a hereditarily finite list, Exercises 3.28 and 3.30 would assure us that there is a hereditarily finite list whose entries are w and the entries of w. When we do our translation, we can let '$w \cup \{w\}$' refer to that list. So our translation says there is a hereditarily finite list x that lists $\emptyset$ and lists $w \cup \{w\}$ whenever it lists w. We can use induction and Theorem 3.7 to show that such an x will list every number.

If Axioms 5.1–5.6 imply that some rank is not a number, then, by the second interpretability result from the previous section, Axioms 3.1–3.3 imply that some hereditarily finite list lists every number. Since this contradicts Theorem 3.17, it would mean that Axioms 3.1–3.3 are inconsistent. If you believe these axioms are consistent, you can think of our argument as a *reductio ad absurdum*. Here is the sequence of entailments, starting with the assumption to be reduced to absurdity.

Axioms 5.1–5.6 prove that some rank is not a number

⇓

Axioms 5.1–5.6 prove that the numbers form a set

⇓

Axioms 3.1–3.3 prove that the numbers form a hereditarily finite list

⇓

Axioms 3.1–3.3 are inconsistent.

We conclude: if Axioms 3.1–3.3 are consistent (or, equivalently, if Axioms 5.1–5.6 are consistent), then Axioms 5.1–5.6 do not imply that there are any ranks other than the numbers $\emptyset$, $S\emptyset$, $SS\emptyset$, ... If we want a rank that is not a number, we will have to adopt a new axiom. We are going to adopt an axiom that gives a rank ω that has all the numbers as its members.

Definition 5.14 α IMMEDIATELY PRECEDES γ if and only if $\alpha \in \gamma$ but for no β do we have $\alpha \in \beta \in \gamma$. A rank that precedes γ is said to be a PREDECESSOR of γ. A rank that immediately precedes γ is said to be γ's IMMEDIATE PREDECESSOR.

Exercise 5.28 *Confirm that a rank can have at most one immediate predecessor. (Note, too, that we already know of a rank with no immediate predecessor.)*

Exercise 5.29 *Confirm that every number other than $\emptyset$ has an immediate predecessor.*

Axiom 5.7 There is a non-empty rank with no immediate predecessor.

Since $\emptyset$ is the only number with no immediate predecessor, Axiom 5.7 implies that some rank is not a number and, so, is independent of the other set axioms (if our first three list axioms are consistent).

Definition 5.15 Let ω be the first rank that satisfies Axiom 5.7

Theorem 5.16 *Every number is a member of ω.*

Proof Exercise 5.29 implies that ω is not a number. Now just apply the argument I gave a few paragraphs ago about ranks that are not numbers. □

Theorem 5.17 *Every member of ω is a number.*

Proof Let β be the first member of ω that is not a number. Then every member of β is a number. Furthermore, since β is a rank that is not a number, every number is a member of β. Given any predecessor n of β, we will have: $n \in Sn \in \beta$. So β is a non-empty rank with no immediate predecessor, contradicting the minimality of ω. □

The members of ω are the numbers $\emptyset$, $S\emptyset$, $SS\emptyset$, ... According to Definition 5.10, the members of $V(\omega)$ are the sets outranked by ω. So we have:

$$x \in V(\omega) \iff \rho(x) \in \omega \iff \rho(x) \text{ is a number.}$$

Just as in the world of lists, something is a number if and only if it is an entry or, as we now say, a member of a number. So:

$$\rho(x) \text{ is a number} \iff \rho(x) \in n \text{ for some number } n.$$

That is, $\rho(x)$ is a number if and only if x is outranked by a number. Since the members of $V(n)$ are the sets outranked by n, we conclude:

$$x \in V(\omega) \iff x \in V(n) \text{ for some number } n.$$

By Definition 5.10, Theorem 5.11, and Exercises 5.8 and 5.27,

$$V(0) = \emptyset$$

$$V(Sn) = PV(n).$$

So, just as in Chap. 3, we can show that the sets $V(n)$ are the super-numbers $\emptyset$, $P\emptyset$, $PP\emptyset$, ... If we say that a hereditarily finite set is any member of a super-number, the members of $V(\omega)$ are the hereditarily finite sets. This confirms the following theorem.

Theorem 5.18 *The hereditarily finite sets form a set.*

For brevity's sake, we say that Axioms 3.1, 3.2, 3.3, and 4.1 are "the list axioms", while Axioms 5.1–5.7 are "the set axioms". We have just confirmed that the translation of Axiom 4.1 is derivable from the set axioms. We already know that the translations of Axioms 3.1–3.3 are derivable. So, if ϕ is a sentence in the language of list theory that follows from the list axioms, the translation of ϕ into the language of set theory will follow from the set axioms. The list axioms, then, are inconsistent only if the set axioms are. Any evidence for the consistency of the set axioms is evidence for the consistency of the list axioms.

Results from Chap. 4 let us establish the converse. To translate statements about sets into statements about lists, we read "set" as "Zermelian list" and "member of" as "entry of". Theorem 4.2 translates Axiom 5.1. Exercises 4.20–4.23 translate Axioms 5.2–5.5. As for Axioms 5.6 and 5.7, note two things. First, every rank is an entry of a rank. Second, ω is a rank with entries and each entry of ω is listed by an entry of ω. We conclude: if ϕ is a sentence in the language of set theory that follows from the set axioms, then the translation of ϕ into the language of list theory will follow from the list axioms. The set axioms are inconsistent only if the list axioms are.

Any evidence for the consistency of the list axioms is evidence for the consistency of the set axioms. Indeed, given our earlier result, we see that the list axioms and the set axioms are equiconsistent: the list axioms are consistent if and only the set axioms are.

5.6 $V(\omega + \omega)$ and Beyond

We have seen that our set axioms all come out true in the universe of Zermelian lists. All the mathematics you are likely to encounter in an undergraduate mathematics course can be reconstructed within this universe. If, however, you want an even more robust set theory, a set theory that makes demands on its models not satisfied by the Zermelian universe, you should consider adding axioms that give you more ranks. As we discussed in Sect. 5.3, ranks are the fuel that drives the production of sets.

What is missing from the universe of Zermelian lists? $\omega + \omega$, if it existed, would list all the entries of all the higher ranks

$$\omega + 0, \omega + 1, \omega + 2, \ldots$$

just as ω lists all the entries of all the numbers

$$0, 1, 2, \ldots.$$

$\omega + \omega$ would, in fact, be the list of all Zermelian ranks. As a transitive set of ranks, $\omega + \omega$ would, by Theorem 5.9, itself be a rank. So, if $\omega + \omega$ were Zermelian, it would list itself. But this cannot happen, because we know Theorem 5.3 is true in the universe of Zermelian lists.[4] So $\omega + \omega$ is missing from the Zermelian universe and it does not follow from our set axioms that there are any ranks other than the numbers $\emptyset, S\emptyset, SS\emptyset, \ldots$ and the higher ranks $\omega, S\omega, SS\omega, \ldots$[5] Summarizing the argument:

- PREMISE: Our set axioms are all true in the universe of Zermelian lists.
- PREMISE: Anything implied by our set axioms will be true wherever those axioms themselves are true.
- PREMISE: In the universe of Zermelian lists, there is no list of all Zermelian ranks.
- CONCLUSION: Our set axioms do not imply that there is a set of such ranks.

---

[4] Similar reasoning yields the Burali-Forti paradox in some versions of set theory. See Burali-Forti [1]; English translation in van Heijenoort [10], pp. 104–111.

[5] OK, there is another alternative. Maybe our set axioms are inconsistent and the whole notion of a universe of Zermelian lists is fundamentally incoherent. A good reason to believe this wrong is that some profound minds have thought deeply about the Zermelian universe and have detected no incoherence. A prominent mathematician who thought he *did* detect incoherence was HERMANN WEYL (1885–1955), who believed classical set theory was "permeated by the poison of contradiction". (See Weyl [11], p. 23; English translation in Weyl [12], p. 32.) For an argument that Weyl should be taken seriously in general, but not in this case, see Pollard [8].

This means it would strengthen our set theory to suppose that, say, there is more than one non-empty rank with no immediate predecessor. This would give us $\omega + \omega$. It would also give us a set $V(\omega + \omega)$ in which our set axioms all come out true: what had served as the universe of all sets would become a set.

Exercise 5.30 *Consider the set theory consisting of Axioms 5.1–5.6 and a version of Axiom 5.7 asserting that there are at least two non-empty ranks with no immediate predecessors. Describe a rank whose existence does not follow from these axioms.*

The preceding exercise might leave us wishing for a less piecemeal approach. No sooner do we escape the Zermelian universe than we discover a rank missing from our *new* universe. Could we manage a more dramatic gesture that gives us lots of ranks all at once and doesn't leave any gaps that are too obvious? Here is an idea. If it existed, the rank $\omega + \omega$ would be well-ordered by the membership relation. Recall, from Sect. 3.6, that a well-ordering is a transitive, irreflexive, connected, well-founded relation. The "less than" relation on the natural numbers is an example. It is transitive

$$m < n < p \quad \text{only if} \quad m < p$$

irreflexive

$$m \not< m$$

and connected

$$m = n \quad \text{or} \quad m < n \quad \text{or} \quad n < m.$$

As for well-foundedness, if you pick any natural numbers, one of them will be less than all the others. (No matter what numbers you pick, there will be a "bottom" one.) The membership relation has these same properties inside of ω and would have them inside of $\omega + \omega$. Well-ordered by membership, $\omega + \omega$ would look like two copies of ω (two copies of the natural numbers) placed one after another. The numbers

$$\emptyset, S\emptyset, SS\emptyset, \ldots$$

would form the first copy of ω, while the higher ranks

$$\omega, S\omega, SS\omega, \ldots$$

would form the second. Now, whether or not $\omega + \omega$ exists, we can re-arrange the natural numbers themselves so that they too look like two copies of ω. For example, we can let the even numbers $0, 2, 4, \ldots$ be the first copy of ω, to be followed by the odd numbers $1, 3, 5, \ldots$ forming the second copy. The natural numbers would then be arranged in the same way as the Zermelian ranks (the numbers n plus the higher ranks $\omega + n$). This suggests the following association of numbers with ranks.

$$f(2n) = n$$

$$f(2n + 1) = \omega + n.$$

The function f associates numbers with ranks as follows.

$$
\begin{array}{ccccccc}
0 & 2 & 4 & \ldots & 1 & 3 & 5 & \ldots \\
\downarrow & \downarrow & \downarrow & & \downarrow & \downarrow & \downarrow \\
0 & 1 & 2 & \ldots & \omega + 0 & \omega + 1 & \omega + 2 & \ldots
\end{array}
$$

Or, if you prefer:

$$
\begin{array}{cccccccc}
0 & 1 & 2 & 3 & 4 & 5 & \ldots \\
\downarrow & \downarrow & \downarrow & \downarrow & \downarrow & \downarrow \\
0 & \omega + 0 & 1 & \omega + 1 & 2 & \omega + 2 & \ldots
\end{array}
$$

Depending on how you look at it, the function f shows us how to make the natural numbers look like the Zermelian ranks or it shows us how to make the Zermelian ranks look like the natural numbers.

How will this help us to get lots of new ranks? Here is a proposition that would allow us to use functions like f to generate ranks.

Proposition 5.1 *Whenever there is a function that takes us from the members of a set to some ranks, those ranks themselves form a set.*

Our function f takes us from the members of the set ω to the Zermelian ranks. So Proposition 5.1 would give us a set whose members are the Zermelian ranks. Proposition 5.1 would give us $\omega + \omega$. Here is another way of thinking about it: if the members of a set, ordered by some relation, exhibit the same structure as some ranks, ordered by membership, then those ranks form a set. Since we can make the members of ω look like two copies of ω and the membership relation makes the Zermelian ranks look like two copies of ω, those ranks form a set. Alternatively, if you can make some ranks look like the members of a set, ordered by some relation, then, again, those ranks form a set. Since we can rearrange the Zermelian ranks so that they look like the members of the set ω, the Zermelian ranks form a set.

Exercise 5.31 *If $\omega \cdot 2$ ("omega times two") looks like two copies of ω and $\omega \cdot n$ looks like n copies of ω, what does ω^2 ("omega squared") look like? Use Proposition 5.1 to show that ω^2 exists.*

Here is one way to motivate Proposition 5.1. If we abstract from everything but order, then the even natural numbers in their standard order look just like all the natural numbers in their standard order.

$$
\begin{array}{l}
0\ 2\ 4\ 6\ 8\ldots \\
0\ 1\ 2\ 3\ 4\ldots
\end{array}
$$

We might say that these two structures are instances or tokens of the same type, the same ORDER TYPE. Each type of *well-ordering* is known as an ORDINAL. Now if we start talking freely about ordinals, someone might reasonably ask us whether these ordinals are supposed to be objects in the universe of sets and, if so, what objects. Structures that look like the members of our rank ω, structures that look like the natural numbers in their standard order, are traditionally said to be of order type ω: their ordinal is ω. Perhaps we should identify the ordinal ω with our rank ω. Perhaps we should, quite generally, identify ordinals with ranks. In the absence of Proposition 5.1, this would lead immediately to problems. We want every well-ordering to be a well-ordering *of some type*. We want there to be an ordinal for every well-ordering. The Zermelian universe $V(\omega + \omega)$ features well-ordered structures of type $\omega + \omega$ but no *rank $\omega + \omega$*. So our set axioms will leave us unable to prove that every well-ordering has an ordinal if we identify ordinals with ranks. Proposition 5.1 is a way of responding to this problem. Given Proposition 5.1, $\omega + \omega$ exists *because* it is the order type of a well-ordered set—that is, because we can make the members of the set ω look like the Zermelian ranks.[6]

Chapters 3–5 have given you three big doses of set theory. You ought to know, however, that you have only just begun to learn about a majestic, audacious mathematical construction that ignited the imagination of some of the most profound minds of the last century. You should explore further on your own.[7] Our next two chapters, however, are going to stray from the set theoretic mainstream. We begin with Frege's uncanny demonstration that arithmetic is a stew consisting almost entirely of ideas and methods drawn from pure logic.

5.7 Solutions of Odd-Numbered Exercises

5.1 $\emptyset$ outranks a set x if and only if some member of $\emptyset$ outranks each member of x. But $\emptyset$ has no members. A set x will outrank $\emptyset$ if and only if some member of x outranks each member of $\emptyset$. If $x \neq \emptyset$, then, by Axiom 5.4, x has members. Any member of x will outrank each member of $\emptyset$—again, because $\emptyset$ has no members.

5.3 By Exercise 5.1, $\emptyset$ does not outrank itself (because it outranks nothing). Suppose no member of x outranks itself. Then no member of x outranks each member of x. So, by Axiom 5.1, x does not outrank itself.

5.5 By Definition 5.6, $\rho(x)$ is a rank. Theorem 5.8 applies to every rank α. So $\rho(\rho(x)) = \rho(x)$.

---

[6] Proposition 5.1 is a version of the Replacement Axiom of ABRAHAM FRAENKEL (1891–1965) and THORALF SKOLEM (1887–1963). See Fraenkel [2], p. 231, and van Heijenoort [10], p. 297. For an especially clear demonstration that Replacement supplies every well-ordered set with an ordinal, see Kunen [5], p. 17. For an overview of the history, see Kanamori [4].

[7] You might start with Kunen [5] for the mathematics and Hallett [3] for the history.

5.7 If $\beta \in \alpha \subseteq \beta$, then $\beta \in \beta$—contrary to Theorem 5.3. So $\beta \notin \alpha$ and, since $\alpha \neq \beta$, Theorem 5.7 guarantees that $\alpha \in \beta$.

5.9 Suppose x is a set of ranks and $\alpha \in x$. By Exercise 5.2, $\rho(\alpha) \in \rho(x)$ and, hence, by Theorem 5.8, $\alpha \in \rho(x)$. We conclude that $x \subseteq \rho(x)$.

5.11 Axiom 5.1 says that my lamp outranks my lamp if and only if some member of my lamp outranks each member of my lamp. Both halves of the biconditional are true: my lamp does outrank my lamp and, furthermore, my lamp's only member outranks my lamp's only member (since my lamp's only member is my lamp). Axiom 5.2 says: if my lamp is of bounded rank, then there is a set whose only member is my lamp. Well, my lamp *is* of bounded rank and is itself a set whose only member is my lamp. Axiom 5.3 is false because the only set in the universe (my lamp) has a member (itself). Axiom 5.4 is true because there is only one set in the universe and, in the absence of distinct sets, there cannot be distinct sets with the same members. Axiom 5.5 is true because our one rank has itself as its only member.

5.13 Pick sets x, y. Theorem 5.7 lets us assume that $\rho(x)$ either precedes $\rho(y)$ or *is* $\rho(y)$. Axiom 5.6 lets us pick a rank α preceded by $\rho(y)$. By Theorem 5.4, $\rho(x)$ precedes α. So the sets x, y are collectively of bounded rank and, hence, by Axiom 5.2, form a set.

5.15 Suppose $z \in y \in x$. By Exercise 5.2 and Theorem 5.2, x outranks z. We conclude: the members of x's members are of bounded rank and, so, we can apply Axiom 5.2.

5.17 Suppose $\beta \in (\alpha \cup \{\alpha\})$. Then either $\beta \in \alpha$ or $\beta = \alpha$. By Theorems 5.3 and 5.4, $\alpha \notin \beta$. We conclude: if α precedes β, then β does not precede $\alpha + 1$. But is $\alpha + 1$ a rank? According to Theorem 5.9, the answer will be "yes" if $\alpha \cup \{\alpha\}$ is a transitive set of ranks. It is a set of ranks because α is both a rank and (Axiom 5.5) a set of ranks. As for transitivity, suppose $\gamma \in \beta \in (\alpha \cup \{\alpha\})$. Case 1: $\beta \in \alpha$. By Theorem 5.4, $\gamma \in \alpha$. Case 2: $\beta = \alpha$. Again, $\gamma \in \alpha$. So, in each case, $\gamma \in (\alpha \cup \{\alpha\})$.

5.19 Suppose $x \in V(\alpha + 1)$. Then $\rho(x) \in (\alpha + 1)$ and, hence, either $\rho(x) \in \alpha$ or $\rho(x) = \alpha$. We want to show that $x \subseteq V(\alpha)$ (since it then follows that $x \in PV(\alpha)$). Suppose $y \in x$. By Exercise 5.2, $\rho(y) \in \rho(x)$ and, hence, by Theorem 5.4, $\rho(y) \in \alpha$. That is, $y \in V(\alpha)$.

5.21 Theorem 5.4 and Definition 5.10 let us reason as follows

$$x \in V(\alpha) \implies \rho(x) \in \alpha \in \beta \implies \rho(x) \in \beta \implies x \in V(\beta).$$

5.23 According to Theorem 5.12, $\rho(V(\alpha))$ is the first rank β such that $V(\alpha) \subseteq V(\beta)$. So, since $V(\alpha) \subseteq V(\alpha)$, $\alpha \notin \rho(V(\alpha))$. On the other hand, suppose $\rho(V(\alpha)) \in \alpha$. Then, by Exercise 5.22, $V(\rho(V(\alpha))) \in V(\alpha)$. But, by Theorem 5.12, $V(\alpha) \subseteq$

$V(\rho(V(\alpha)))$. So $V(\rho(V(\alpha))) \in V(\rho(V(\alpha)))$—contrary to Exercises 5.2 and 5.3. Now apply Theorem 5.7.

5.25 Suppose $v \in w \in z$. Since w is an M-set, so is v. Since w outranks v but does not outrank y, v does not outrank y. So $v \in z$.

5.27 Suppose y is transitive. We want to show that $y \cup \{y\}$ is transitive. Suppose $w \in x \in (y \cup \{y\})$. Then either $x \in y$ or $x = y$. In each case, $w \in y$ and, hence, $w \in (y \cup \{y\})$. Since $y \cup \{y\}$ is transitive and $y \in (y \cup \{y\})$, every precursor of y is a member of $y \cup \{y\}$. That is, $Sy \subseteq (y \cup \{y\})$. On the other hand, $(y \cup \{y\}) \subseteq Sy$ since y and all its members are precursors of y. Now apply Axiom 5.4.

5.29 Suppose $n \in k \in Sn$. By Exercise 5.27, either $n \in k = n$ or $n \in k \in n$—contrary to Theorems 5.3 and 5.4. So the immediate predecessor of Sn is n. Every number is either $\emptyset$ or a successor. So every non-zero number has an immediate predecessor.

5.31 ω^2 looks like ω copies of ω. Here is one way to rearrange the members of $\omega \backslash \{0\}$ to look like that. Let the odd numbers be the first copy of ω. Multiply each odd number by 2 to get the second copy. Multiply *those* numbers by 2 to get the third copy. And so on. We end up with the following structure.

$$
\begin{array}{cccccc}
1 & 3 & 5 & 7 & 9 & 11 \ldots \\
2 & 6 & 10 & 14 & 18 & 22 \ldots \\
4 & 12 & 20 & 28 & 36 & 44 \ldots \\
8 & 24 & 40 & 56 & 72 & 88 \ldots \\
\vdots & \vdots & \vdots & \vdots & \vdots & \ddots
\end{array}
$$

The nth copy of ω is the result of multiplying each odd number by 2^{n-1}. The factorization $(2m - 1) \cdot 2^{n-1}$ tells us to look in column m on row n. For example,

$$36 = 9 \cdot 2^2 = (2 \cdot 5 - 1) \cdot 2^{3-1}.$$

So 36 appears in column 5 on row 3. Since every positive integer is the product of an odd number and a power of 2, we are not leaving out any members of $\omega \backslash \{0\}$. Here is another way to think about it. Write down the odd numbers and the powers of 2 first.

$$
\begin{array}{l}
1\ 3\ 5\ 7\ 9\ 11 \ldots \\
2 \\
4 \\
8 \\
\vdots
\end{array}
$$

Treat the odd numbers $3, 5, 7, 9, 11, \ldots$ as column labels and the powers $2, 4, 8, \ldots$ as row labels. Fill in the table by putting $j \cdot k$ in column j on row k.

References

1. Burali-Forti, C. (1897). Una questione sui numeri transfiniti. *Rendiconti del Circolo matematico di Palermo, 11,* 154–164.
2. Fraenkel, A. (1922). Zu den Grundlagen der Cantor-Zermeloschen Mengenlehre. *Mathematische Annalen, 86,* 230–237.
3. Hallett, M. (1984). *Cantorian set theory and limitation of size.* Oxford: Clarendon Press.
4. Kanamori, A. (2012). In praise of replacement. *Bulletin of Symbolic Logic, 18,* 46–90.
5. Kunen, K. (1980). *Set theory: An introduction to independence proofs.* Amsterdam: North-Holland.
6. Pollard, S. (1990). *Philosophical introduction to set theory.* Notre Dame IN: University of Notre Dame Press.
7. Pollard, S. (1992). Choice again. *Philosophical Studies, 66,* 285–296.
8. Pollard, S. (2005). Property is prior to set: Fichte and Weyl. In G. Sica (Ed.), *Essays on the foundations of mathematics and logic* (pp. 209–226). Monza, Italy: Polimetrica.
9. Van Aken, J. (1986). Axioms for the set theoretic hierarchy. *Journal of Symbolic Logic, 51,* 992–1004.
10. van Heijenoort, J. (Ed.). (1967). *From Frege to Gödel.* Cambridge MA: Harvard University Press.
11. Weyl, H. (1918). *Das Kontinuum.* Leipzig: Veit.
12. Weyl, H. (1994). *The continuum.* New York: Dover.

Chapter 6
Frege Arithmetic

6.1 The Language of **FA**

If you ask to see a formalization of arithmetic, a mathematician who knows about such things will probably point you toward Peano Arithmetic. We will now consider an alternative formalization of number theory derived from the work of Gottlob Frege. We will also consider an unsuccessful Fregean approach to the foundations of set theory. As with our treatment of **PA** in Chap. 2, we will not present a formal logic for our new version of number theory. Our proofs will be informal. Be assured, however, that Frege himself provided a fully formalized version of the underlying logic.[1]

The language of FREGE ARITHMETIC (**FA**) has the following vocabulary.

1. Two CONNECTIVES: '$-$' ("not"), '$\rightarrow$' ("if …then").
2. A QUANTIFIER: '$\forall$' ("for all").
3. The IDENTITY symbol: '$=$'.
4. One FUNCTION symbol: '$\#$' ("number of").
5. Infinitely many OBJECT VARIABLES: 'w', 'x', 'y', 'z', 'w_1', 'x_1', 'y_1', 'z_1', 'w_2', 'x_2', 'y_2', 'z_2', …
6. Infinitely many PROPERTY VARIABLES: 'F', 'G', 'H', 'F_1', 'G_1', 'H_1', …
7. Infinitely many RELATION VARIABLES: 'P', 'Q', 'R', 'P_1', 'Q_1', 'R_1', …
8. Two PARENTHESES: '(', ')'.

The OBJECT TERMS of **FA** are the object variables and any expression $\ulcorner \# \Gamma \urcorner$ where Γ is a property variable. The idea is that $\# F$ is the number of things that have property F. If F were the property of being a prefecture of Japan, then $\# F$ would be the number of Japanese prefectures: that is, $\# F$ would be 47. As the expression

---

[1] Informality will not be our only departure from Frege. So be warned: this chapter does not pretend to offer a close reading of Frege's own exposition. It draws freely on later reconstructions such as the one in George and Velleman [7]. For Frege's own version, see [4] and [5]. For some helpful commentary, see Boolos and Heck [1].

S. Pollard, *A Mathematical Prelude to the Philosophy of Mathematics*,
DOI: 10.1007/978-3-319-05816-0_6,
© Springer International Publishing Switzerland 2014

"object term" suggests, numbers are classified as *objects*. # is a function that assigns an object (a number) to each property.

We recursively define the FORMULAS of **FA** as follows.

9. If α and β are object terms, then $\ulcorner \alpha = \beta \urcorner$ is a formula.
10. If α is an object term and Γ is a property variable, then $\ulcorner \Gamma \alpha \urcorner$ is a formula.
11. If α and β are object terms and Σ is a relation variable, then $\ulcorner \Sigma \alpha \beta \urcorner$ is a formula.
12. If ϕ and ψ are formulas, then so are $\ulcorner -\phi \urcorner$ and $\ulcorner (\phi \rightarrow \psi) \urcorner$.
13. If μ is a variable and ϕ is a formula, then $\ulcorner \forall \mu \ \phi \urcorner$ is a formula.

Property variables range over properties. 'Fx' says that object x has property F. Relation variables range over binary (two-place) relations. 'Rxy' says that object x stands in relation R to object y.

We define FREE and BOUND occurrences of variables just as we did in Chap. 2. A SENTENCE is a formula with no free occurrences of variables. We now stipulate:

$$(\psi \wedge \chi) \iff -(\psi \rightarrow -\chi)$$
$$(\psi \leftrightarrow \chi) \iff ((\psi \rightarrow \chi) \wedge (\chi \rightarrow \psi))$$
$$(\psi \vee \chi) \iff (-\psi \rightarrow \chi)$$
$$\exists \mu \ \phi \iff -\forall \mu - \phi.$$

'$\wedge$' represents CONJUNCTION ("and"), '$\leftrightarrow$' represents the BICONDITIONAL ("if and only if"), while '$\vee$' represents DISJUNCTION ("or"). '$\exists$' is the EXISTENTIAL QUANTIFIER ("there is").

Exercise 6.1 *Translate the following sentences of* **FA** *into English.*

$$\forall F \forall G (\forall x (Fx \leftrightarrow Gx) \rightarrow \#F = \#G).$$

$$\forall F \forall G (\#F = \#G \rightarrow \forall x (Fx \leftrightarrow Gx)).$$

$$\forall F \exists G (-\forall x (Fx \leftrightarrow Gx) \wedge \#F = \#G).$$

$$\exists F \forall G (\#F = \#G \rightarrow -\exists x \ Gx).$$

$$\forall R (\exists x \forall y \ Rxy \rightarrow \forall y \exists x \ Rxy).$$

We use recursion to introduce NUMERICAL QUANTIFIERS $\ulcorner \exists_n \urcorner$ ("there are exactly n") as follows.

Definition 6.1

$$\exists_0 x \ \phi x \iff -\exists x \ \phi x.$$
$$\exists_{n+1} x \ \phi x \iff \exists x (\phi x \wedge \exists_n y (\phi y \wedge x \neq y)).$$

'$\exists_0 x \ Fx$' says no object has property F: $-\exists x \ Fx$. In English: "It is false that there is an object with property F". '$\exists_1 x \ Fx$' says exactly one object has property F:

$$\exists x (Fx \wedge \exists_0 y (Fy \wedge x \neq y)).$$

In English: "There is an object x that has property F and no object other than x has property F". '$\exists_2 x \ Fx$' says exactly two objects have property F:

$$\exists x (Fx \wedge \exists_1 y (Fy \wedge x \neq y)).$$

In English: "There is an object x that has property F and exactly one object other than x has property F". And so on.

The following definition will be particularly useful.

Definition 6.2

$$\forall F \forall G (F \approx G \leftrightarrow \exists R (\forall x (Fx \to \exists_1 y (Rxy \wedge Gy)) \wedge \forall y (Gy \to \exists_1 x (Rxy \wedge Fx)))).$$

Suppose $F \approx G$ and let R be a relation of the sort required by the definition. Say that the objects with properties F and G are, respectively, the F-objects and the G-objects. Then R assigns to each F-object exactly one G-object. Furthermore, each G-object is assigned by R to exactly one F-object. So R PAIRS F-objects with G-objects. We also say that R is a PAIRING.

Suppose, for example, that F is the property of being a New England state capital, while G is the property of being a New England state. Let R be a relation that holds between a city and a state when the former is the capital of the latter. Then R assigns to each New England state capital exactly one New England state, while each New England state is assigned by R to exactly one New England state capital. Indeed, R forms the following pairing.

Hartford	$\xrightarrow{\text{is the capital of}}$	Connecticut
Augusta	$\xrightarrow{\text{is the capital of}}$	Maine
Boston	$\xrightarrow{\text{is the capital of}}$	Massachusetts
Concord	$\xrightarrow{\text{is the capital of}}$	New Hampshire
Providence	$\xrightarrow{\text{is the capital of}}$	Rhode Island
Montpelier	$\xrightarrow{\text{is the capital of}}$	Vermont

From the existence of such a pairing we can infer that the F-objects and the G-objects are EQUINUMEROUS, that is, the same in number.

6.2 The Axioms of FA

We will now make the unsurprising assumption that the F-objects and the G-objects are the same in number if and only if the number of F-objects is the same as the number of G-objects. This assumption is known as HUME'S PRINCIPLE.[2]

$$\forall F \forall G (F \approx G \leftrightarrow \#F = \#G).$$

We also adopt infinitely many COMPREHENSION axioms for properties. These are the sentences of the form

$$\ulcorner \forall \bar{\mu} \exists \Gamma \forall \alpha (\Gamma \alpha \leftrightarrow \phi) \urcorner$$

where α is an object variable, Γ is a property variable, ϕ is a formula of FA with no free occurrences of Γ, and $\forall \bar{\mu}$ is a string of universal quantifiers that bind all the free occurrences of variables in the remainder of the sentence. Here is an example of a comprehension axiom:

$$\forall G \forall y \exists F \forall x (Fx \leftrightarrow (Gx \wedge x \neq y)).$$

This says: given any property G and object y, there is a property F that applies, exactly, to those G-objects that are not y. If G were the property of being a U.S. president and y were the current president, then you could think of F as the property of being a past president.

Exercise 6.2 *Can you think of a highly undesirable comprehension axiom we would obtain if we allowed Γ to occur free in ϕ?*

Exercise 6.3 *The language of FA features a symbol '=' expressing the relation of identity. Show that this symbol is eliminable by offering a definition of identity. Naturally, '=' should not occur in your definition. You might recall from a logic class that two fundamental properties of identity are: (1) everything is identical to itself and (2) identical things have the same properties. Verify that your definition is adequate by showing that it entails these two properties (Students of modern philosophy might find it helpful to review Leibniz.).*

Here is another one of our comprehension axioms:

$$\exists F \forall x (Fx \leftrightarrow x \neq x).$$

It says there is a property that applies to nothing. Let $\mathfrak{N}$ be such a property.

---

[2] In Sect. 63 of [3] (English translation in [6]), Frege quotes the following remark by DAVID HUME (1711–1776): "When two numbers are so combined as that one has always an unit answering to every unit of the other, we pronounce them equal".

Exercise 6.4 *Use comprehension to show that there is a property $\mathfrak{U}$ that applies to exactly one object (By the way, how many objects are we acquainted with at this point?). Use Hume's principle to show that $\#\mathfrak{N}$ is not the only number in the universe.*

Exercise 6.5 *Use comprehension to show that there is a property $\mathfrak{D}$ that applies to exactly two objects. Use Hume's principle to show that $\#\mathfrak{N}$ and $\#\mathfrak{U}$ are not the only numbers in the universe. How many numbers are there?*

We conclude this section by adopting infinitely many COMPREHENSION axioms for relations. These are the sentences of the form

$$\ulcorner\forall\bar{\mu}\exists\Sigma\forall\alpha\forall\beta(\Sigma\alpha\beta \leftrightarrow \phi)\urcorner$$

where α, β are object variables, Σ is a relation variable, ϕ is a formula of **FA** with no free occurrences of Σ, and $\forall\bar{\mu}$ is a string of universal quantifiers that bind all the free occurrences of variables in the remainder of the sentence. Here is an example of a comprehension axiom of this sort:

$$\exists R\forall x\forall y(Rxy \leftrightarrow \exists F(Fx \wedge Fy)).$$

This sentence says there is a relation R that holds between any objects x and y that share at least one property.

Exercise 6.6 *Show that if R behaves in the way just described, then $\forall x\forall y Rxy$.*

Exercise 6.7 *Show: $\forall F(\#F = \#\mathfrak{N} \leftrightarrow \exists_0 x\ Fx)$.*

Exercise 6.8 *Show: $\forall F(\#F = \#\mathfrak{U} \leftrightarrow \exists_1 x\ Fx)$.*

6.3 Some Number Theory

Let $0 = \#\mathfrak{N}$ where $\mathfrak{N}$ is, as in the previous section, a property that applies to nothing:

$$\forall x(\mathfrak{N}x \leftrightarrow x \neq x).$$

This should sound reasonable: 0 is the number of objects that are not identical to themselves; 0 is the number of objects that have a property nothing could possibly have. This definition certainly seems less contrived than our stipulation in Chap. 3 that 0 is the blank list. To express the whole business in the language of lists: it seems less contrived to say that 0 is the number of entries of the blank list than to stipulate that 0 *is* the blank list.

Let $1 = \#\mathfrak{U}$ where $\mathfrak{U}$ is a property whose existence is guaranteed by the following comprehension axiom:

$$\exists F\forall x(Fx \leftrightarrow x = 0).$$

That is, $\mathfrak{U}$ is a property that applies only to 0:

$$\forall x(\mathfrak{U}x \leftrightarrow x = 0).$$

Since 0 is the only object that has property $\mathfrak{U}$, the number of $\mathfrak{U}$-objects is, indeed, one. 1 is the number of objects identical to 0. Again, this seems less contrived than our earlier stipulation that 1 is the list whose only entry is the blank list.

We know, of course, that 1 is the first number after 0: that 1 *immediately succeeds* 0. We now want to give a formal definition of immediate succession using the language of FA. Since our definition will be a bit complicated, we are going to sneak up on it one little step at a time. We begin with a theorem of FA that may appear entirely pointless:

$$0 = 0 \wedge \forall y(y \neq y \leftrightarrow (y = 0 \wedge y \neq 0)).$$

From this we infer the following:

$$\mathfrak{U}0 \wedge \forall y(\mathfrak{N}y \leftrightarrow (\mathfrak{U}y \wedge y \neq 0)).$$

So:

$$\exists x(\mathfrak{U}x \wedge \forall y(\mathfrak{N}y \leftrightarrow (\mathfrak{U}y \wedge y \neq x))).$$

Adding a bit more information about $\mathfrak{N}$ and $\mathfrak{U}$:

$$0 = \#\mathfrak{N} \wedge 1 = \#\mathfrak{U} \wedge \exists x(\mathfrak{U}x \wedge \forall y(\mathfrak{N}y \leftrightarrow (\mathfrak{U}y \wedge y \neq x))).$$

We conclude:

$$\exists F \exists G(0 = \#F \wedge 1 = \#G \wedge \exists x(Gx \wedge \forall y(Fy \leftrightarrow (Gy \wedge y \neq x)))).$$

Since we derived this sentence from a theorem of FA, we know it too is a theorem of FA. Our theorem says there are properties F and G with the following characteristics: 0 is the number of F-objects, 1 is the number of G-objects, and, for some G-object x, the F-objects are exactly the G-objects distinct from x. That is, we can obtain the F-objects by deleting one G-object or, in other words, the number of G-objects SUCCEEDS the number of F-objects (I am going to save ink by leaving off the 'immediately' in 'immediately succeeds'.). In symbols: $\sigma\#F\#G$ ('σ' is the Greek letter sigma.). Since $\#F = 0$ and $\#G = 1$, we have just confirmed that 1 succeeds 0: $\sigma01$. We have also provided an analysis of this relation of succession. We now enshrine this analysis in a general definition of succession.

Definition 6.3

$$\forall w \forall z(\sigma wz \leftrightarrow \exists F \exists G(w = \#F \wedge z = \#G \wedge \exists x(Gx \wedge \forall y(Fy \leftrightarrow (Gy \wedge y \neq x))))).$$

z succeeds w if and only if w is the number of F-objects (for some property F), z is the number of G-objects (for some property G), and we can obtain the F-objects by deleting one of the G-objects.

Exercise 6.9 *Show:* $\exists_0 w\ \sigma w 0$.

Theorem 6.1 $\forall w \forall z_1 \forall z_2((\sigma w z_1 \wedge \sigma w z_2) \rightarrow z_1 = z_2)$.

Proof Let F_1, G_1 be properties certifying that $\sigma w z_1$, while F_2, G_2 are properties certifying that $\sigma w z_2$. That is,

$$w = \#F_1 \wedge z_1 = \#G_1 \wedge \exists x_1(G_1 x_1 \wedge \forall y(F_1 y \leftrightarrow (G_1 y \wedge y \neq x_1)))$$

and

$$w = \#F_2 \wedge z_2 = \#G_2 \wedge \exists x_2(G_2 x_2 \wedge \forall y(F_2 y \leftrightarrow (G_2 y \wedge y \neq x_2))).$$

We need to show that $\#G_1 = \#G_2$ and this, in turn, requires us to show that $G_1 \approx G_2$. Note, first, that $\#F_1 = \#F_2$ since both are identical to w. So Hume's principle allows us to pick a relation R that pairs F_1-objects with F_2-objects. Comprehension gives us another pairing P with the following useful feature:

$$\forall x \forall y(Pxy \leftrightarrow (F_1 x \wedge F_2 y \wedge Rxy)).$$

That is, P ignores everything but F_1-objects and F_2-objects, pairing the former with the latter. Now pick a G_1-object x_1 and a G_2-object x_2 that behave as indicated above. That is, first, the F_1-objects are exactly the G_1-objects distinct from x_1 and, second, the F_2-objects are exactly the G_2-objects distinct from x_2. We will use these materials to identify a relation that pairs G_1-objects and G_2-objects. Comprehension provides a relation Q that behaves as follows:

$$\forall x \forall y(Qxy \leftrightarrow (Pxy \vee (x = x_1 \wedge y = x_2))).$$

We want to show that Q is the desired pairing. Suppose x is a G_1-object. If $x = x_1$, then Qxx_2. Suppose $x \neq x_1$. Then x is an F_1-object and, hence, we can pick an F_2-object y such that Pxy. So Qxy. Since every F_2-object is a G_2-object, y is a G_2-object. We conclude that Q assigns a G_2-object to every G_1-object. We still need to show that this assignment is unique. Suppose:

$$G_1 x, \quad Qxz, \quad Qxz'.$$

Case 1: $x = x_1$. Then x is not an F_1-object and, hence, it is not the case that Pxy for any object y. So Qxy only if $y = x_2$. This means that z and z' are the same G_2-object, namely, x_2. Case 2: $x \neq x_1$. Then Pxz and Pxz'. So z and z' are the same F_2-object and, hence, are the same G_2-object. We conclude: given any G_1-object x, there is a unique G_2-object y such that Qxy. Similar reasoning shows: given any

G_2-object y, there is a unique G_1-object x such that Qxy. We conclude: $G_1 \approx G_2$. So, by Hume's principle, $\#G_1 = \#G_2$.

Theorem 6.2 $\forall w \forall z_1 \forall z_2 ((\sigma w_1 z \wedge \sigma w_2 z) \rightarrow w_1 = w_2)$.

Proof Let F_1, G_1 be properties certifying that $\sigma w_1 z$, while F_2, G_2 are properties certifying that $\sigma w_2 z$. That is,

$$w_1 = \#F_1 \wedge z = \#G_1 \wedge \exists x_1 (G_1 x_1 \wedge \forall y (F_1 y \leftrightarrow (G_1 y \wedge y \neq x_1)))$$

and

$$w_2 = \#F_2 \wedge z = \#G_2 \wedge \exists x_2 (G_2 x_2 \wedge \forall y (F_2 y \leftrightarrow (G_2 y \wedge y \neq x_2))).$$

We need to show that $F_1 \approx F_2$. Since $\#G_1 = \#G_2$, Hume's principle allows us to pick a relation R that pairs G_1-objects with G_2-objects. We also pick a G_1-object x_1 and a G_2-object x_2 as indicated above. The G_1-objects are the F_1-objects plus x_1, while the G_2-objects are the F_2-objects plus x_2. If R pairs x_1 with x_2, then R already pairs F_1-objects with F_2-objects (since they are what is left when we have disposed of x_1 and x_2). Suppose, on the other hand, that $Rx_1 x_2'$ and $Rx_1' x_2$ where $x_2' \neq x_2$. Then $x_1' \neq x_1$. Furthermore, x_1' is an F_1-object, while x_2' is an F_2-object. We want to make an adjustment: we want x_1' and x_2' to be paired with one another. Comprehension provides a relation Q that behaves as follows:

$$\forall x \forall y (Qxy \leftrightarrow ((x = x_1' \wedge y = x_2') \vee (x \neq x_1' \wedge y \neq x_2' \wedge Rxy))).$$

We want to show that Q is the desired pairing. Suppose x is an F_1-object. If $x = x_1'$, then Qxx_2'. Suppose $x \neq x_1'$. Pick a y such that Rxy. Then $y \neq x_2$ and, hence, y is an F_2-object. If $y = x_2'$, then $x = x_1$, which is impossible since x_1 is not an F_1-object. So $y \neq x_2'$ and, hence, Qxy. We conclude that Q assigns an F_2-object to every F_1-object. We still need to show that this assignment is unique. Suppose:

$$F_1 x, \quad Qxz, \quad Qxz'.$$

Case 1: $x = x_1'$. Then Qxy only if $y = x_2'$. So z and z' are the same F_2-object, namely, x_2'. Case 2: $x \neq x_1'$. Then Rxz and Rxz' and, hence, $z = z'$. We conclude: given any F_1-object x, there is a unique F_2-object y such that Qxy. Similar reasoning shows: given any F_2-object y, there is a unique F_1-object x such that Qxy. We conclude: $F_1 \approx F_2$. So, by Hume's principle, $\#F_1 = \#F_2$.

Definition 6.4 A property F is σ-CLOSED if and only if:

$$\forall x \forall y ((Fx \wedge \sigma xy) \rightarrow Fy).$$

A property is σ-closed property if and only if all the objects that have it pass it on to all their successors.

Definition 6.5 $x \leq y$ if and only if y has every σ-closed property that x has.

$x \leq y$ if and only if y is one of the objects tireless immortal beings will encounter if they start with x and track down all the successors of everything they encounter.

Theorem 6.3 $\leq$ *is reflexive and transitive.*

Proof Reflexivity: $x \leq x$ because x has every σ-closed property that x has. Transitivity: if y has every σ-closed property that x has and z has every σ-closed property that y has, then z has every σ-closed property that x has.

Theorem 6.4 $\forall x \forall y (\sigma x y \rightarrow x \leq y)$.

Proof If $\sigma x y$, then x gives each of its σ-closed properties to y.

Theorem 6.5 $\forall z (z \leq 0 \rightarrow z = 0)$.

Proof Suppose $z \leq 0$. Then 0 has every σ-closed property that z has. Exercise 6.9 implies that the property "not identical to 0" is σ-closed. That is,

$$\forall x \forall y ((x \neq 0 \wedge \sigma x y) \rightarrow y \neq 0)$$

since

$$\forall x \forall y (\sigma x y \rightarrow y \neq 0).$$

So $z \neq 0$ only if $0 \neq 0$. Since we are rewarded with an absurdity if $z \neq 0$, we conclude that $z = 0$.

Definition 6.6 $\forall x \forall y (x < y \leftrightarrow (x \leq y \wedge x \neq y))$.

Theorem 6.6 $\forall w \forall x \forall y ((\sigma x y \wedge w < y) \rightarrow w \leq x)$.

Proof Suppose F is a σ-closed property that applies to w. We want to show that it also applies to x. Use comprehension to pick a property G such that

$$\forall z (Gz \leftrightarrow (Fz \wedge z \neq y)).$$

Note that $w \neq y$ since $w < y$. So Gw since we just supposed that Fw. y has every σ-closed property that w has. So G can be σ-closed only if $y \neq y$. Clearly, G is *not* σ-closed. So we can pick z_1, z_2 such that $\sigma z_1 z_2$ even though G applies to z_1 and not z_2. Since Gz_1, we have: Fz_1 and $z_1 \neq y$. Since *not* Gz_2, we have: Fz_2 only if $z_2 = y$. Since F is σ-closed and applies to z_1, it applies to z_2. So $z_2 = y$. Hence, by Theorem 6.2, $z_1 = x$, since $\sigma z_1 y$ and $\sigma x y$. So Fx, as desired. We conclude that x has every σ-closed property that w has.

Exercise 6.10 *Show:* $\forall x \forall y \forall z ((\sigma x y \wedge x < z) \rightarrow y \leq z)$. *Hint: suppose F is a σ-closed property that applies to y; consider the following comprehension axiom:*

$$\exists G \forall w (Gw \leftrightarrow (Fw \vee w = x)).$$

Definition 6.7 x is a NATURAL NUMBER if and only if $0 \leq x$.

Natural numbers are objects that are doomed whenever 0 catches a disease that is transmitted to successors. Less fancifully, the natural numbers are the objects possessing every σ-closed property that applies to 0.

Theorem 6.7 *0 is a natural number.*

Exercise 6.11 *Show: every object that succeeds a natural number is a natural number.*

Theorem 6.8 *σ-closed properties that apply to 0 apply to every natural number.*

Say that something is a NUMBER if and only if it is the number of things that have some property. That is, z is a number if and only if $z = \#G$ for some property G. Then every natural number is a number. To confirm this, first recall that $0 = \#\mathfrak{N}$. So 0 is a number. Now look back at Definition 6.3 to confirm that σwz only if $z = \#G$ for some property G. So the property of being a number is σ-closed. Now just apply Theorem 6.8.

Exercise 6.12 *Use induction (Theorem 6.8) to show that no natural number succeeds itself.*

Theorem 6.9 *If x and y are natural numbers, then σxy only if $x < y$.*

Proof Just apply Theorem 6.4 and Exercise 6.12.

Exercise 6.13 *Use induction to show that every natural number other than 0 succeeds a natural number.*

Exercise 6.14 *Use induction and Exercise 6.10 to show: if x, y are natural numbers, then either $x \leq y$ or $y \leq x$.*

Theorem 6.10 *If x and y are natural numbers and $y \leq x$, then $-\sigma xy$.*

Proof Exercise 6.9 guarantees that the theorem holds when $y = 0$. Suppose σyz. Our argument is inductive: we want to show that z satisfies the theorem if y does. Suppose $z \leq x$. We want to show that $-\sigma xz$. Exercise 6.13 says we need only consider two cases. First: $x = 0$. Then, by Theorem 6.5, $z = 0$. So, by Exercise 6.9 (or 6.12), $-\sigma xz$. Second: x succeeds a natural number. Suppose σwx. If $z = x$, then Exercise 6.12 guarantees that $-\sigma xz$. Suppose $z \neq x$. Then $z < x$ and, hence, by Theorem 6.6, $z \leq w$. By Theorem 6.4, $y \leq z$. So, by Theorem 6.3, $y \leq w$. Our inductive hypothesis is that y satisfies the theorem. So $-\sigma wy$ and, hence, $x \neq y$. By Theorem 6.2, $-\sigma xz$.

Exercise 6.15 *Use induction to show that if x and y are natural numbers, then $x \leq y$ and $y \leq x$ only if $x = y$. (You might find Theorems 6.3, 6.4, 6.5, 6.6, and 6.10 useful.)*

Exercise 6.16 *Show that if x, y are natural numbers, then σxy only if*

$$\forall w(w \leq x \leftrightarrow w < y).$$

(Again, you might find Theorems 6.6 and 6.10 useful.)

Theorem 6.11 *Suppose x_1, x_2 are natural numbers and $\sigma x_1 x_2$. If*

$$\forall y(F_1 y \leftrightarrow y \leq x_1),$$

$$\forall y(F_2 y \leftrightarrow y \leq x_2),$$

then $\sigma \# F_1 \# F_2$.

Proof According to Exercise 6.16,

$$\forall y(y \leq x_1 \leftrightarrow (y \leq x_2 \wedge y \neq x_2)).$$

So, since $x_2 \leq x_2$,

$$\exists x(x \leq x_2 \wedge \forall y(y \leq x_1 \leftrightarrow (y \leq x_2 \wedge y \neq x))).$$

So:

$$\exists x(F_2 x \wedge \forall y(F_1 y \leftrightarrow (F_2 y \wedge y \neq x))).$$

So:

$$\exists F \exists G(\# F_1 = \# F \wedge \# F_2 = \# G \wedge \exists x(Gx \wedge \forall y(Fy \leftrightarrow (Gy \wedge y \neq x)))).$$

That is, $\sigma \# F_1 \# F_2$.

Exercise 6.17 *Show that $\forall y(Fy \leftrightarrow y < 0)$ only if $\# F = 0$.*

Exercise 6.18 *Suppose x_1, x_2 are natural numbers and $\sigma x_1 x_2$. Further, suppose $\# F_1 = x_1$ whenever F_1 is a property satisfying the following condition:*

$$\forall y(F_1 y \leftrightarrow y < x_1).$$

Show that if

$$\forall y(F_2 y \leftrightarrow y < x_2),$$

then $\# F_2 = x_2$ (You might show that $\sigma x_1 \# F_2$ and then apply Theorem 6.1.).

Theorem 6.12 *If x is a natural number and $\forall y(Fy \leftrightarrow y < x)$, then $\# F = x$.*

Proof Note that the previous two exercises provide an inductive argument for this theorem.

We now know that each natural number n is the number of objects less than n. Suppose $x < m$ where m is a natural number. By Theorem 6.2 and Exercise 6.13, the property of *not* being a natural number is σ-closed (since, otherwise, some natural number would have a predecessor that both is and is not a natural number). So, since m has every σ-closed property that x has, there is no way for x not to be a natural number. We conclude that each natural number n is the number of *natural numbers* less than n.

Exercise 6.19 *Show that every natural number has a successor. (According to Exercise 6.13, you only need to consider two cases: 0 and a natural number x_2 that succeeds a natural number x_1. The first case is already done: we showed that σ01. As for the second case, you might take a look at Theorem 6.11.)*

Suppose **N** is a property that applies to all and only the natural numbers, while $\mathbf{N}^+$ is a property that applies to all and only the positive (i.e., non-zero) natural numbers. Then the successor relation σ is a pairing between the **N**-objects and the $\mathbf{N}^+$-objects. So $\mathbf{N} \approx \mathbf{N}^+$ and, hence, by Hume's principle, $\#\mathbf{N} = \#\mathbf{N}^+$. On the other hand,

$$\exists x(\mathbf{N}x \wedge \forall y(\mathbf{N}^+y \leftrightarrow (\mathbf{N}y \wedge y \neq x))).$$

So $\#\mathbf{N}$ succeeds $\#\mathbf{N}^+$ and, hence, $\#\mathbf{N}$ succeeds itself. Exercise 6.12 lets us conclude that $\#\mathbf{N}$ is not a natural number. Every natural number is a number, but not every number is a natural number. In particular, the number of natural numbers is not a natural number. The number of natural numbers cannot be measured by any finite number: not a million, not a billion, not

$$10^{10^{10}}.$$

OK, that is no surprise. There are infinitely many natural numbers. What *is* surprising, even uncanny, is the way this result emerges from a lot of logical manipulations and a few applications of one innocent looking observation about the conditions under which numbers are the same: the number of F-objects is the same as the number of G-objects if and only the F-objects and the G-objects are equal in number. Something that makes Hume's principle look all the more innocent, and the emergence of infinitely many natural numbers all the more uncanny, is the definability of "equal in number" in our background *logic*. Granted, our logic does not provide a definition of "number of" that lets us eliminate the operator '#'. '#' is not eliminable in the way 'σ' is. So our development of arithmetic requires more than logic. But it does not require *much* more. When we get inside some of the proofs given above, it is striking how much of the work consists of logical manipulations of logical vocabulary— manipulations that make us feel like we have wandered into a logic class. FA treats number theory as applied logic. Just as applied mathematics can look an awful lot like mathematics, applied logic can look an awful lot like logic. At the very least, FA gives number theory a strongly logical flavor.

6.4 FA Interprets PA

We still need to figure out how much number theory FA provides. We can give a precise answer to this question if we introduce a beefier version of Peano Arithmetic (PA). Let the vocabulary of SECOND- ORDER PEANO ARITHMETIC consist of the vocabulary of PA and the whole vocabulary of FA except for the function symbol '#'. Second-order PA includes all the comprehension axioms in this vocabulary. The remaining axioms of second-order PA are the same as those of PA except that we drop the infinitely many induction axioms and replace them with the single sentence:

$$\forall F((F0 \land \forall x(Fx \rightarrow FSx)) \rightarrow \forall x \, Fx).$$

To decipher the expression 'FSx' you need to remember that 'S' is a function symbol in the language of PA. 'Sx' is a term referring to the successor of x. The formula 'FSx' says that the successor of x has property F. Our induction principle says that if 0 has property F and the successor of x has property F whenever x does, then everything in the universe of PA (that is, every natural number) has property F.

FA provides at least as much number theory as second-order PA because, as we will confirm, second-order PA is interpretable in FA. Though we will not prove it, the converse is also true.[3] So FA provides exactly as much number theory as second-order PA because the two theories are mutually interpretable.

When we offer FA translations of PA axioms, we will feel free to use items of *defined* vocabulary such as '0' and '$\leq$'. This is legitimate because our definitions allow us to eliminate all occurrences of the defined expressions. That means we are always able to recover the official version. For example, if we offer $\phi(0)$ as our translation of a PA axiom and someone objects that '0' is not part of the official vocabulary of FA, we can offer the alternative translation

$$\exists F(\forall x(Fx \leftrightarrow x \neq x) \land \phi(\#F)).$$

A faithful translation will preserve correct inferences: if ψ follows from $\phi_1, ..., \phi_n$, then the translation of ψ will follow from the translations of $\phi_1, ..., \phi_n$. A translation that behaves otherwise would not even teach us about relative consistency (Recall §5.4.). Here is an example of what we expect from our translations. Since ($\phi(0) \land \psi(0)$) follows from $\phi(0)$ and $\psi(0)$, our translation of the former sentence should follow from our translations of the latter sentences. So

$$\exists F(\forall x(Fx \leftrightarrow x \neq x) \land (\phi(\#F) \land \psi(\#F)))$$

should follow from

$$\exists F(\forall x(Fx \leftrightarrow x \neq x) \land \phi(\#F))$$

and
$$\exists G(\forall x(Gx \leftrightarrow x \neq x) \land \psi(\#G)).$$

Well, it *does* follow. If $\phi(\#F)$ and $\psi(\#G)$ and, furthermore, F and G are properties that apply to nothing, then, by Exercise 6.7, $\#F = \#G$ and, hence, $(\phi(\#F) \land \psi(\#F))$. This is hardly a proof that our translations will *always* preserve correct inferences. It is just an indication of what such a proof would involve (You might try to work out what a real proof would look like.).

The following definition will let us deal smoothly with occurrences of 'S' in axioms of PA. The definition is justified because each natural number has one (Exercise 6.19) and only one (Theorem 6.1) successor.

Definition 6.8 $\forall x(0 \leq x \rightarrow \forall y(y = Sx \leftrightarrow \sigma xy)).$

We will take advantage of this definition in our translations of the PA axioms S1 and S2. Inside of PA, we pretend that everything in the universe is a natural number. In FA, we allow for objects (such as the number of natural numbers) that are not natural numbers. So, when PA says "all x" our FA translation says "all natural numbers x"—that is, according to Definition 6.7, "all x such that $0 \leq x$". We translate S1 and S2 as follows.
$$\forall x(0 \leq x \rightarrow 0 \neq Sx)$$

$$\forall x\forall y((0 \leq x \land 0 \leq y) \rightarrow (Sx = Sy \rightarrow x = y))$$

Exercise 6.9 and Theorem 6.2 let us verify that these are theorems of FA.

What about the PA induction axiom given above? Is its translation an FA theorem? Well, Theorem 6.8 is the induction principle of FA. We can capture it in a single sentence of FA:

$$\forall F((F0 \land \forall x\forall y((Fx \land \sigma xy) \rightarrow Fy)) \rightarrow \forall x(0 \leq x \rightarrow Fx)).$$

This says that if 0 has property F and F is σ-closed, then every natural number has property F. Here is our translation of the PA induction axiom:

$$\forall F((F0 \land \forall x((0 \leq x \land Fx) \rightarrow FSx)) \rightarrow \forall x(0 \leq x \rightarrow Fx)).$$

Is this a theorem of FA? To confirm that it is, suppose $F0$ and

$$\forall x((0 \leq x \land Fx) \rightarrow FSx).$$

Then, by Exercise 6.11,

$$\forall x((0 \leq x \land Fx) \rightarrow (0 \leq Sx \land FSx)).$$

Use comprehension to pick a property G such that

$$\forall x(Gx \leftrightarrow (0 \le x \wedge Fx)).$$

Then $\forall x(Gx \rightarrow GSx)$ and, hence,

$$\forall x \forall y((Gx \wedge \sigma xy) \rightarrow Gy).$$

That is, G is σ-closed. Furthermore, by Theorem 6.7, $G0$. So, by the FA induction axiom, $\forall x(0 \le x \rightarrow Gx)$ and, hence, $\forall x(0 \le x \rightarrow Fx)$. We conclude that our translation of the PA induction axiom is a theorem of FA.

Our next job is to confirm that translations of A1, A2, M1, and M2 are theorems of FA. This will take a bit of work. Our job would be easier if comprehension supplied us immediately with a relation R that behaves as follows:

$$\forall w \forall y_1 \forall y_2(Ry_1y_2 \leftrightarrow ((y_1 = 0 \wedge y_2 = w) \vee \exists x_1(\sigma x_1 y_1 \wedge \exists x_2(Rx_1x_2 \wedge \sigma x_2 y_2)))).$$

Once we pick a w, R is supposed to be the relation that holds between two natural numbers when the second is exactly w more than the first. That is:

$$Ry_1y_2 \iff y_2 = y_1 + w.$$

Our formula considers two possibilities. First, $y_1 = 0$. Note that y_2 is exactly w more than 0 if and only if $y_2 = w$:

$$y_2 = 0 + w \iff y_2 = w.$$

That explains the clause

$$(y_1 = 0 \wedge y_2 = w).$$

The second possibility is that y_1 succeeds x_1. That is, $y_1 = x_1 + 1$. Note that y_2 is exactly w more than $x_1 + 1$ if and only if $y_2 - 1$ is exactly w more than x_1:

$$y_2 = (x_1 + 1) + w = (x_1 + w) + 1 \iff y_2 - 1 = x_1 + w.$$

Or, letting $y_2 = x_2 + 1$:

$$x_2 + 1 = (x_1 + 1) + w = (x_1 + w) + 1 \iff x_2 = x_1 + w.$$

That is, y_2 is exactly w more than y_1 if and only if x_2 is exactly w more than x_1. More briefly, Ry_1y_2 if and only if Rx_1x_2. That explains the clause

$$\exists x_1(\sigma x_1 y_1 \wedge \exists x_2(Rx_1x_2 \wedge \sigma x_2 y_2))).$$

This is all very elegant.

Unfortunately, we have a problem. The formula at the start of the last paragraph is not a comprehension axiom: it violates the requirement that 'R' not appear free on the right side of a comprehension biconditional when R is the relation on the left side. Nonetheless, we can use our formula to define a useful notion. If, for a given natural number w, a relation R satisfies the formula

$$Ry_1y_2 \leftrightarrow ((y_1 = 0 \wedge y_2 = w) \vee \exists x_1(\sigma x_1 y_1 \wedge \exists x_2(Rx_1x_2 \wedge \sigma x_2 y_2)))$$

for all natural numbers y_1, y_2, then we say that R is a "plus-w" relation. We need to show that there is a plus-w relation for each natural number w. We accomplish this by induction in the following exercise and theorem.

Exercise 6.20 *Show that there is a plus-0 relation.*

Theorem 6.13 *If w is a natural number and Q is a plus-w relation, then there is a plus-Sw relation.*

Proof Comprehension supplies a relation R that behaves as follows:

$$Ry_1y_3 \leftrightarrow \exists y_2(Qy_1y_2 \wedge \sigma y_2y_3).$$

R is to be our plus-Sw or plus-$(w + 1)$ relation. The idea is:

$$y_1 + (w + 1) = y_3 \leftrightarrow (y_1 + w) + 1 = y_3.$$

Since Q is a plus-w relation:

$$Qy_1y_2 \leftrightarrow ((y_1 = 0 \wedge y_2 = w) \vee \exists x_1(\sigma x_1 y_1 \wedge \exists x_2(Qx_1x_2 \wedge \sigma x_2y_2))).$$

That is:

$$Qy_1y_2 \leftrightarrow ((y_1 = 0 \wedge y_2 = w) \vee \exists x_1(\sigma x_1 y_1 \wedge Rx_1y_2)).$$

We want to show that R is a plus-Sw relation. That is,

$$Ry_1y_3 \leftrightarrow ((y_1 = 0 \wedge y_3 = Sw) \vee \exists x_1(\sigma x_1 y_1 \wedge \exists y_2(Rx_1y_2 \wedge \sigma y_2y_3)))$$

whenever y_1 and y_3 are natural numbers. To verify the left-right half of the biconditional, we assume that Ry_1y_3. This lets us pick a y_2 such that Qy_1y_2 and σy_2y_3. Then:

$$(y_1 = 0 \wedge y_2 = w) \vee \exists x_1(\sigma x_1 y_1 \wedge Rx_1y_2).$$

By Theorem 6.1,

$$(y_1 = 0 \wedge y_2 = w) \rightarrow (y_1 = 0 \wedge y_3 = Sw)$$

since σwSw. Furthermore:

$$\exists x_1(\sigma x_1 y_1 \wedge Rx_1y_2) \rightarrow \exists x_1(\sigma x_1 y_1 \wedge (Rx_1y_2 \wedge \sigma y_2y_3)).$$

But:

$$\exists x_1(\sigma x_1 y_1 \wedge (R x_1 y_2 \wedge \sigma y_2 y_3)) \rightarrow \exists x_1(\sigma x_1 y_1 \wedge \exists y_2(R x_1 y_2 \wedge \sigma y_2 y_3)).$$

So:

$$R y_1 y_3 \rightarrow ((y_1 = 0 \wedge y_3 = Sw) \vee \exists x_1(\sigma x_1 y_1 \wedge \exists y_2(R x_1 y_2 \wedge \sigma y_2 y_3))).$$

Now we need to show the converse. We consider two cases. First:

$$y_1 = 0 \wedge y_3 = Sw.$$

Note that $Q0w$ and, hence, $\exists y_2(Q0y_2 \wedge \sigma y_2 Sw)$. That is, $R0Sw$. But then $Ry_1 y_3$. Here is the second case:

$$\exists x_1(\sigma x_1 y_1 \wedge \exists y_2(R x_1 y_2 \wedge \sigma y_2 y_3)).$$

Pick such a y_2. Then $Q y_1 y_2$ since $\exists x_1(\sigma x_1 y_1 \wedge R x_1 y_2)$. So $\exists y_2(Q y_1 y_2 \wedge \sigma y_2 y_3)$ and, hence, $R y_1 y_3$. We conclude that R is a plus-Sw relation.

Exercise 6.21 *Suppose R is a plus-w relation. Use induction to show that, for every natural number y_1, there is a natural number y_2 such that $R y_1 y_2$.*

Exercise 6.22 *Suppose R is a plus-w relation. Use induction to show that, for any natural numbers y_1, y_2, y_3, if $R y_1 y_2$ and $R y_1 y_3$, then $y_2 = y_3$.*

Exercise 6.23 *Suppose Q and R are plus-w relations. Use induction to show that, for any natural numbers y_1, y_2, if $Q y_1 y_2$, then $R y_1 y_2$.*

Theorem 6.14 *If y_1 and y_2 are natural numbers and R is a plus-w relation, then $R y_1 y_2$ if and only if $R S y_1 S y_2$.*

Proof Note, first, that

$$R S y_1 S y_2 \leftrightarrow ((S y_1 = 0 \wedge S y_2 = w) \vee \exists x_1(\sigma x_1 S y_1 \wedge \exists x_2(R x_1 x_2 \wedge \sigma x_2 S y_2))).$$

So, by Exercise 6.9,

$$R S y_1 S y_2 \leftrightarrow \exists x_1(\sigma x_1 S y_1 \wedge \exists x_2(R x_1 x_2 \wedge \sigma x_2 S y_2)).$$

Note that $\sigma y_1 S y_1$ and $\sigma y_2 S y_2$. So, by Theorem 6.2,

$$\exists x_1(\sigma x_1 S y_1 \wedge \exists x_2(R x_1 x_2 \wedge \sigma x_2 S y_2)) \leftrightarrow R y_1 y_2.$$

Exercise 6.24 *Suppose R is a plus-0 relation. Use induction to show that, for every natural number x, Rxx.*

Definition 6.9 If w and y are natural numbers, we let $y + w$ be the unique natural number that every plus-w relation associates with y.

Now that we have introduced '+' into the language of FA, we can show that straightforward translations of the PA addition axioms are theorems of FA. Here is our translation of A1.

Theorem 6.15 $\forall x(0 \leq x \rightarrow (x + 0) = x)$.

Proof $x + 0$ is the unique natural number that every plus-0 relation associates with x. But, according to Exercise 6.24, that number is x.

Theorem 6.16 $\forall x \forall y((0 \leq x \wedge 0 \leq y) \rightarrow (Sx + y) = S(x + y))$.

Proof Let R be a plus-y relation. Then $Rx(x + y)$ and, hence, by Theorem 6.14, $RSxS(x + y)$. But $RSx(Sx + y)$ since $Sx + y$ is the unique natural number that every plus-y relation associates with Sx. So, by Exercise 6.22, $(Sx + y) = S(x + y)$.

Here is our translation of A2.

Theorem 6.17 $\forall x \forall y((0 \leq x \wedge 0 \leq y) \rightarrow (x + Sy) = S(x + y))$.

Proof We use induction. y is the unique natural number that every plus-y relation associates with 0. That is, $(0 + y) = y$. Sy is the unique natural number that every plus-Sy relation associates with 0. That is, $(0 + Sy) = Sy$. So

$$(0 + Sy) = Sy = S(0 + y)$$

and, hence, the theorem holds when $x = 0$. As an inductive hypothesis, suppose $(x + Sy) = S(x + y)$. Then, with some help from Theorem 6.16, we can reason as follows:
$$(Sx + Sy) = S(x + Sy) = SS(x + y) = S(Sx + y).$$

Exercise 6.25 *Using M1 and M2 from Chap. 2 and our definition of plus-w relations as guides, introduce the notion of "times-w" relations.*

Exercise 6.26 *Show that there is a times-0 relation.*

Exercise 6.27 *Show: if w is a natural number and Q is a times-w relation, then there is a times-Sw relation.*

Exercise 6.28 *Suppose R is a times-w relation. Use induction to show that, for every natural number y_1, there is a natural number y_2 such that Ry_1y_2.*

Exercise 6.29 *Suppose R is a times-w relation. Use induction to show that, for any natural numbers y_1, y_2, y_3, if Ry_1y_2 and Ry_1y_3, then $y_2 = y_3$.*

Exercise 6.30 *Suppose Q and R are times-w relations. Use induction to show that, for any natural numbers y_1, y_2, if Qy_1y_2, then Ry_1y_2.*

If x is a natural number, we can let $x \cdot w$ be the unique natural number that every times-w relation associates with x. Then we can prove the following in **FA**.

$$\forall x (0 \leq x \to (x \cdot 0) = 0)$$

$$\forall x \forall y ((0 \leq x \wedge 0 \leq y) \to (x \cdot Sy) = ((x \cdot y) + x)).$$

That is, we can show that translations of the **PA** axioms M1 and M2 are theorems of **FA**. You might want to fill in the details yourself.

The only remaining axioms of second-order **PA** are the infinitely many comprehension axioms. Consider one with the following form.

$$\forall w \exists F \forall x (Fx \leftrightarrow \phi)$$

Let ϕ^i be the **FA** formula that translates the **PA** formula ϕ. Then

$$\forall w \exists F \forall x (Fx \leftrightarrow \phi^i)$$

is a comprehension axiom of **FA**. This axiom implies

$$\forall w (0 \leq w \to \exists F \forall x (0 \leq x \to (Fx \leftrightarrow \phi^i)))$$

which is our translation of the **PA** comprehension axiom. We conclude that every **PA** comprehension axiom translates into a theorem of **FA**. And our grand conclusion is that second-order **PA** is interpretable in **FA**.

6.5 Extensions: An Epic Failure

We will now consider a modification of **FA**: FREGE ARITHMETIC WITH EXTENSIONS (**FAX**).[4] The vocabulary of **FAX** is the result of dropping '#' from the vocabulary of **FA** and adding the symbols '{', '}', and ':'. The OBJECT TERMS of **FAX** are the object variables and any expression $\ulcorner \{\alpha : \Gamma \alpha\} \urcorner$ where α is an object variable and Γ is a property variable. The idea is that $\{x : Fx\}$ is the EXTENSION of the property F: it is the set of F-objects. If F were the property of being a dog, then $\{x : Fx\}$ would be the set of dogs. As the expression "object term" suggests, extensions are classified as objects. We define the FORMULAS of **FAX** just as we did the formulas of **FA**. We obtain the axioms of **FAX** by dropping Hume's principle and adding BASIC LAW V.[5]

$$[BLV] \quad \forall F \forall G (\{x : Fx\} = \{x : Gx\} \leftrightarrow \forall x (Fx \leftrightarrow Gx)).$$

---

[4] This is a version of Frege's system in [4] and [5]. '**FAX**' is just *our* name for this theory: do not expect anyone else to call it that.

[5] The 'V' is a Roman numeral: BLV is the fifth "basic law" from Frege's *Grundgesetze* ([4], [5]).

The idea is: the set of F-objects is the same as the set of G-objects if and only if the F-objects are the same as the G-objects.

As reasonable as Basic Law V might sound, we will now confirm that **FAX** is inconsistent. Comprehension supplies a property G that behaves as follows:

$$\forall y(Gy \leftrightarrow \exists F(\{x : Fx\} = y \wedge -Fy)).$$

Each G-object is an extension that lacks the very property of which it is the extension. For example, the set of all non-self-identical objects (the set of all objects x such that $x \neq x$) is a G-object because it *is* identical to itself and, hence, lacks the property (non-self-identity) of which it is the extension. The set of all self-identical objects (the set of all objects x such that $x = x$) is *not* a G-object because it is identical to itself and, hence, has the property (self-identity) of which it is the extension.

We are going to consider the set of all G-objects: $\{x : Gx\}$. The instance of comprehension we just cited makes a claim about "any object y". Letting y be $\{x : Gx\}$, we obtain:

$$G\{x : Gx\} \leftrightarrow \exists F(\{x : Fx\} = \{x : Gx\} \wedge -F\{x : Gx\}).$$

Suppose $\{x : Gx\}$ is not a G-object. That is, suppose $-G\{x : Gx\}$. Then:

$$-\exists F(\{x : Fx\} = \{x : Gx\} \wedge -F\{x : Gx\}).$$

Logic students may recognize that this is equivalent to:

$$\forall F(\{x : Fx\} = \{x : Gx\} \rightarrow F\{x : Gx\}).$$

This sentence makes a claim about "any property F". Letting F be G, we obtain:

$$\{x : Gx\} = \{x : Gx\} \rightarrow G\{x : Gx\}.$$

But it is certainly true that $\{x : Gx\} = \{x : Gx\}$. So $G\{x : Gx\}$. That is, $\{x : Gx\}$ *is* a G-object. We got to this point by assuming that $\{x : Gx\}$ is not a G-object. So we have shown:

$$-G\{x : Gx\} \rightarrow G\{x : Gx\}.$$

It follows that $\{x : Gx\}$ really is a G-object (because if it were not, then it *would* be and we would find ourselves in the absurd situation that it both is and is not). This means:

$$\exists F(\{x : Fx\} = \{x : Gx\} \wedge -F\{x : Gx\}).$$

Pick such an F. Then $\{x : Gx\}$ is not an F-object. Furthermore, by Basic Law V:

$$\forall x(Fx \leftrightarrow Gx).$$

In particular:

$$F\{x : Gx\} \leftrightarrow G\{x : Gx\}.$$

So, since $\{x : Gx\}$ is a G-object, it is an F-object. That is, $\{x : Gx\}$ both is and is not an F-object and, so, both is and is not a G-object. We conclude that Basic Law V is an exceptionally plausible-looking *absurdity*.

6.6 The Perils of Abstraction

Frege's plan was to use extensions to define the function # and to do so in a way that allowed him to derive Hume's principle. He would then develop arithmetic much as we did above. The discovery that **FAX** is inconsistent left this project in ruins.[6] Frege's failure to notice the inconsistency of **FAX** may appear more forgivable if we consider the logical kinship of Hume's principle and Basic Law V. We begin with a definition.

Definition 6.10 $\forall F \forall G(F \equiv G \leftrightarrow \forall x(Fx \leftrightarrow Gx)).$

Properties that apply to the same objects are said to be EXTENSIONALLY EQUIV-ALENT: $\equiv$ is the relation of extensional equivalence. Properties that apply to the same number of objects are said to be CARDINALLY EQUIVALENT: $\approx$ is the relation of cardinal equivalence. Both extensional equivalence and cardinal equivalence are, indeed, EQUIVALENCE RELATIONS: that is, transitive, symmetric, and reflexive.[7]

Let us now write '$\mathfrak{X}F$' instead of '$\{x : Fx\}$'. This makes the similarity between Hume's principle and Basic Law V even more evident.

$$\forall F \forall G(\#F = \#G \leftrightarrow F \approx G)$$
$$\forall F \forall G(\mathfrak{X}F = \mathfrak{X}G \leftrightarrow F \equiv G)$$

Propositions of this form are known as ABSTRACTION PRINCIPLES. Hume's principle says that properties have the same cardinality if and only if they are cardinally equivalent. Basic Law V says that properties have the same extension if and only if they are extensionally equivalent. It may not sound too outlandish to say that HP is part of the very "logic" of the concept of cardinality, while BLV is part of the very "logic" of the concept of extension. The formal similarity of HP and BLV tempts us to accord them the same logical status. Of course, we know this is wrong-headed. When you add HP to second-order logic, you get arithmetic. When you add BLV to second-order logic, you get nonsense. It is hard to imagine a more significant difference from a logical perspective.

---

[6] For some of the history, see Quine [8] (http://www.jstor.org/stable/2251464), Rang & Thomas [9], and van Heijenoort [11], pp. 124–128.

[7] TRANSITIVITY: if $F \equiv G$ and $G \equiv H$, then $F \equiv H$. SYMMETRY: if $F \equiv G$, then $G \equiv F$. REFLEXIVITY: $F \equiv F$.

At least one conclusion seems inescapable: if Hume's principle is a logical or conceptual truth, this cannot be simply because of its logical form. Some energetic philosophers have labored hard to identify other distinctive features of Hume's principle that justify assigning it some sort of exalted status. Knowledgeable people disagree about the extent to which these efforts have been crowned with success.[8]

Even if we deny that Hume's principle is a logical truth, we can still be impressed by how much of the heavy lifting in FA is performed by the logical machinery. FA is a departure from the set theoretic mainstream that relies heavily on a powerful version of classical logic. In our next chapter, we will consider an alternative to classical logic itself. This is another way to break free from the prevailing current. First, though, we consider a Fregean theory of extensions that is not known to be absurd.

6.7 Monadic Frege Arithmetic

In FAX, extensions of properties are objects to which properties apply. So it makes sense to ask whether a property does or does not apply to its own extension. This leads to disaster: the property "being the extension of a property that does not apply to its own extension" both does and does not apply to its own extension.

One popular fix is to keep the extensions but lose the properties: we let the variables 'F', 'G', 'H', ...range over sets while the variables 'w', 'x', 'y', ...range over members of those sets. 'Fx' will now mean that object x is a member of set F. It will often be convenient to express this in the more familiar way: '$x \in F$'. We also express *non*-membership in the familiar way: '$x \notin F$'. It is characteristic of sets that they are the same if they have the same members. We enshrine this principle in an axiom (and save a little ink by writing '$\forall F, G$' for '$\forall F \forall G$').

Axiom 6.1 $\forall F, G(F \equiv G \rightarrow F = G)$.

This is one half of BLV with both occurrences of '$\mathfrak{X}$' deleted. Of course, '$\mathfrak{X}F$' no longer makes much sense. Sets do not have extensions, they *are* extensions. The extension of set F would be, if anything, F itself.

In the language of FAX we could ask whether $F\mathfrak{X}F$—that is, whether property F applies to extension $\mathfrak{X}F$. We might now think that expressions such as 'FF' or '$F \in F$' make claims that, true or false, at least make sense. We might want to use such expressions to raise the question of whether set F is a member of itself. Our syntax, however, does not care how we feel about this: 'FF' is not even a formula. So we cannot even ask whether the set of all self-membered sets is self-membered (as we asked, in Chap. 3, whether the list of all lists that do not list themselves lists itself). 'FF' was not a formula in FAX either, but FAX featured a device, the "extension-of" operator, that blasted through the grammatical barrier by converting property terms

---

(such as 'F') into object terms (such as '$\mathcal{X}F$') in a particularly unfortunate way. We are going to try hard to retain the grammatical prohibition on set self-membership. We *will* still use the "number-of" operator to convert set terms (such as 'F') into object terms (such as '$\#F$'). Our syntax will not prevent the cardinality of a set (the number of its members) from being a member of that set. As far as we know, this is harmless.

We are going to deviate from FAX (and, indeed, FA) in one other respect. We are going to jettison the entire apparatus of relation variables and relational comprehension axioms. Our logic is then said to be MONADIC SECOND-ORDER. So we say our new theory is MONADIC FREGE ARITHMETIC (MFA).[9] Object variables are first-order. Set variables are second-order. Set variables are, furthermore, monadic because each applies to only one object variable at a time: 'Fx' is grammatical, but 'Fxy' is not. One reason for going monadic is that it lets us investigate in detail the contribution that the relational machinery made to the construction of arithmetic in FA. We will keep careful track of the new axioms and vocabulary we adopt to compensate for the absence of relation variables.

We do not have to wait long to feel our loss. Look back at Definition 6.2. We used relation variables to define cardinal equivalence. Although that route is no longer open to us, we can still treat '$\approx$' as a defined term. We just use Hume's principle as our definition.

Definition 6.11 $\forall F, G(F \approx G \leftrightarrow \#F = \#G)$.

Since we have not yet said how the function $\#$ behaves, this might not seem very informative. Nonetheless, we can make a little progress with just this much information. First of all, we can show that $\approx$ is an equivalence relation.

Exercise 6.31 $\approx$ *is reflexive, symmetric, and transitive.*

$\approx$ cannot be just *any* equivalence relation, however.

Exercise 6.32 *If all our comprehension axioms are true, then* $\approx$ *cannot be identity:* $\approx$ *cannot be* $=$.[10]

It is an agreeable pastime to construct finite models of comprehension and axiom 6.1. Here is a recipe for doing so. Start with n objects. This yields 2^n sets of objects. To interpret '$\approx$', partition those sets into EQUIVALENCE CLASSES: classes of sets equivalent to one another and to nothing else. When you perform this step, just make sure you put each of your 2^n sets into exactly one equivalence class (You can

---

[9] Yet again, this is just *our* name for the theory.

[10] That is, the function $\#$ cannot be one-to-one:

$$-\forall F, G(\#F = \#G \rightarrow F = G).$$

This is a version of a result known as "Cantor's theorem". GEORG CANTOR (1845–1918) was a pioneer set theorist. The basic problem here is that you cannot use *objects* to supply each *set* of objects with an avatar exclusive to that set—or, rather, you cannot do so while also making all our comprehension axioms true.

put different sets in the same class, but you must not put the same set in different classes.). When you say what the equivalence classes are, you are also saying how you interpret '$\approx$': $F \approx G$ if and only if F and G belong to the same equivalence class. If $\{F_1, F_2, F_3\}$ is one of your equivalence classes, then you are saying that the three sets F_1, F_2, F_3 are equivalent only to one another and each one to itself. To interpret '#', pair off equivalence classes with objects—pairing each equivalence class with exactly one object and each object with no more than one equivalence class. If we pair off equivalence class $\{F_1, F_2, F_3\}$ with object x, we are telling the function # to assign x to each member of $\{F_1, F_2, F_3\}$ and to nothing else: we are saying, first, that $\#F_1 = \#F_2 = \#F_3 = x$ and, second, that # does not assign x to any other sets.

Here is an example. Suppose x is our only object. Then we have two sets of objects: the set of no objects $\emptyset$ and the set of all objects $\{x\}$. This leaves us two ways to form equivalence classes. First, we could put our two sets in the same equivalence class: $\{\emptyset, \{x\}\}$. This makes every set (all two of them) equivalent to every set. Second, we could put our two sets in separate equivalence classes: $\{\emptyset\}$ and $\{\{x\}\}$. This makes each set equivalent only to itself. This latter approach treats $\approx$ as identity and, as you just showed, is not compatible with an interpretation that makes all our comprehension axioms true.[11] The former approach makes $\emptyset$ equivalent to $\{x\}$: $\emptyset \approx \{x\}$. Given the one equivalence class $\{\emptyset, \{x\}\}$ and the one object x, we must pair the former with the latter. That is, we must let # assign x to each member of our one equivalence class: $\#\emptyset = \#\{x\} = x$. This interpretation makes Axiom 6.1 and all instances of comprehension true. Of course, it is hardly the *intended* interpretation since it says that $\emptyset$, a set with no members, is equivalent to $\{x\}$, a set with one member. Our intention is for $\approx$ to behave like cardinal equivalence (Cardinally equivalent sets are sets with the same number of members.).

In our universe of one object and two sets, the only way to get $\approx$ to behave like cardinal equivalence would be to let $\approx$ be the relation of identity: the one set with zero members would then be cardinally equivalent only to itself; the one set with one member would be cardinally equivalent only to itself. But, as just noted, this interpretation does not make all our axioms true. So, in the tiniest universe, we cannot satisfy our axioms by treating $\approx$ as cardinal equivalence. This is just one instance of a general phenomenon: there are no finite models that treat $\approx$ as cardinal equivalence. The reason is simple. In a universe of n objects, sets of objects come in $n + 1$ sizes: 0 through n. So there will not be enough objects to serve as sizes of sets. Axioms that make $\approx$ behave like cardinal equivalence will imply that the universe is infinite. Adding the right sort of information about $\approx$ will let us prove there are infinitely many objects and sets of objects. In FA, we used relation variables to convey this information. We are now trying to do this job without relation variables.

---

[11] Note that you will not be able to complete the remaining step in the recipe I just gave. Why? Given one object and two equivalence classes, you will not be able to pair each equivalence class with exactly one object and each object with no more than one equivalence class. You are facing a crippling shortage of objects.

Given any object z and sets F and G, comprehension and Axiom 6.1 guarantee that there are unique sets $\emptyset$, $\{z\}$, $(F \cup G)$, $(F \setminus G)$, and $(F \cap G)$ satisfying the following conditions.

$$\forall x (x \in \emptyset \leftrightarrow x \neq x)$$
$$\forall x (x \in \{z\} \leftrightarrow x = z)$$
$$\forall x (x \in (F \cup G) \leftrightarrow (x \in F \vee x \in G))$$
$$\forall x (x \in (F \setminus G) \leftrightarrow (x \in F \wedge x \notin G))$$
$$\forall x (x \in (F \cap G) \leftrightarrow (x \in F \wedge x \in G))$$

We stipulate that $0 = \#\emptyset$. We retain Definition 6.3 as our account of the successor relation σ. Given Axiom 6.1, Definition 6.3 is equivalent to the following.[12]

$$\forall w, z (\sigma w z \leftrightarrow \exists G \exists x \in G (w = \#(G \setminus \{x\}) \wedge z = \#G)).$$

Note that

$$\exists G \exists x \in G (w = \#(G \setminus \{x\}) \wedge z = \#G) \leftrightarrow \exists F \exists x \notin F (w = \#F \wedge z = \#(F \cup \{x\}))$$

since, given G and x, we can let F be $G \setminus \{x\}$ and, given F and x, we can let G be $F \cup \{x\}$. We now introduce a successor relation that holds directly between sets.

Definition 6.12 $\forall F, G (S(F, G) \leftrightarrow \exists x \notin F (F \cup \{x\}) \approx G)$.

Exercise 6.33 $\forall w, z (\sigma w z \leftrightarrow \exists F, G (w = \#F \wedge z = \#G \wedge S(F, G)))$.

Exercise 6.34 $\forall F, G, H ((S(F, G) \wedge G \approx H) \rightarrow S(F, H))$.

We define the natural numbers as we did above (though we now refer to sets rather than properties). Comprehension and Axiom 6.1 guarantee that there is exactly one

---

[12] As another ink-saving measure, we write

$$\ulcorner \exists \mu \in \Gamma \ \phi \urcorner$$

instead of

$$\ulcorner \exists \mu (\mu \in \Gamma \wedge \phi) \urcorner.$$

We will also write

$$\ulcorner \forall \mu \in \Gamma \ \phi \urcorner$$

instead of

$$\ulcorner \forall \mu (\mu \in \Gamma \rightarrow \phi) \urcorner.$$

We treat '$\notin$' similarly. For example, we write

$$\ulcorner \exists \mu \notin \Gamma \ \phi \urcorner$$

instead of

$$\ulcorner \exists \mu (\mu \notin \Gamma \wedge \phi) \urcorner.$$

set ω whose members are the natural numbers. It quickly follows that $0 \in \omega$ and that $\omega \subseteq F$ whenever F is σ-closed and $0 \in F$ (As in Definition 5.7, $x \subseteq y$, that is, x is a SUBSET of y, if and only if every member of x is a member of y.).

Exercise 6.35 ω *is σ-closed.*

We say that a set F is FINITE when $\#F$ is a natural number.

Definition 6.13 $\forall F(\mathfrak{F}(F) \leftrightarrow \#F \in \omega)$.

The empty set $\emptyset$ is finite. Adding one member to a finite set yields a finite set. Sets equivalent to finite sets are finite.

Exercise 6.36 $\mathfrak{F}(\emptyset)$.

Exercise 6.37 $\forall F(\mathfrak{F}(F) \rightarrow \forall x\, \mathfrak{F}(F \cup \{x\}))$.

Exercise 6.38 $\forall F, G((F \approx G \wedge \mathfrak{F}(F)) \rightarrow \mathfrak{F}(G))$.

The following result will be a powerful tool for proving facts about finite sets.

Theorem 6.18 *If $\phi(\emptyset)$ and*

$$\forall F(\phi(F) \rightarrow \forall x\, \phi(F \cup \{x\}))$$

$$\forall F, G((\phi(F) \wedge F \approx G) \rightarrow \phi(G))$$

then $\forall G(\mathfrak{F}(G) \rightarrow \phi(G))$.

Proof This is actually a meta-theorem (a theorem about the existence of theorems) governing arbitrary formulas ϕ. Comprehension lets us pick a set H such that

$$\forall x(x \in H \leftrightarrow \exists F(\phi(F) \wedge \#F = x)).$$

We are going to use induction to show that every natural number is a member of H. $0 \in H$, since $\phi(\emptyset)$ and $\#\emptyset = 0$. Suppose $w \in H$ and σwz. Exercise 6.33 lets us pick F_1, G, and x such that $w = \#F_1, z = \#G, x \notin F_1$, and $(F_1 \cup \{x\}) \approx G$. Since $w \in H$ we may pick an F_2 such that $\phi(F_2)$ and $\#F_2 = w$. By Definition 6.11, $F_2 \approx F_1$ and, hence, $\phi(F_1)$. So $\phi(F_1 \cup \{x\})$ and, hence, $\phi(G)$. So $z \in H$. That is, H is σ-closed. So $\omega \subseteq H$. Suppose $\mathfrak{F}(G_1)$. By Definition 6.13, $\#G_1 \in \omega$ and, hence, $\#G_1 \in H$. So we can pick an F_3 such that $\phi(F_3)$ and $\#F_3 = \#G_1$. By Definition 6.11, $F_3 \approx G_1$ and, hence, $\phi(G_1)$.

H is a PROPER subset of G ($H \subset G$) if and only if $H \subseteq G$ but $H \neq G$. We say that $F \prec G$ when F is equivalent to a proper subset of G.

Definition 6.14 $\forall F, G(F \prec G \leftrightarrow \exists H(F \approx H \wedge H \subset G))$.

Exercise 6.39 $\forall F, G(F \subset G \rightarrow F \prec G)$.

Exercise 6.40 $\forall F \; F \not\approx \emptyset$.

Exercise 6.41 $\forall F(F \neq \emptyset \rightarrow \emptyset \prec F)$.

Definition 6.15 $\forall G(\mathfrak{F}_{\prec}(G) \leftrightarrow (\mathfrak{F}(G) \wedge \forall F(F \prec G \rightarrow \mathfrak{F}(F))))$.

Exercise 6.42 $\mathfrak{F}_{\prec}(\emptyset)$.

Exercise 6.43 $\forall F, G, H((F \approx G \wedge G \prec H) \rightarrow F \prec H)$.

You just showed: if F is equivalent to G and G is equivalent to a proper subset of H, then F is equivalent to a proper subset of H. What if we switch this around a bit? What if F is equivalent to a proper subset of G and G is equivalent to H? That is,

$$F \prec G \approx H$$

instead of

$$F \approx G \prec H.$$

Does it *then* follow that F is equivalent to a proper subset of H ($F \prec H$)? This does not follow from our current axioms. In the tiny model we discussed above, $\{0\} \approx \emptyset \subset \{0\}$ and, hence, $\{0\} \prec \{0\} \approx \emptyset$. But $\{0\} \not\approx \emptyset$. ($\{0\}$ is not equivalent to a proper subset of $\emptyset$ because $\emptyset$ has no proper subsets.) Since our variation on Exercise 6.34 is a reasonable, powerful, but underivable claim about cardinal equivalence (our intended interpretation of '$\approx$'), we adopt it as an axiom. (You might take a moment to note, though, how far we have gotten with just comprehension and Axiom 6.1.)

Axiom 6.2 $\forall F, G, H((F \prec G \wedge G \approx H) \rightarrow F \prec H)$.

Exercise 6.44 $\forall F(F \approx \emptyset \rightarrow F = \emptyset)$.

Exercise 6.45 $\prec$ *is transitive.*

Exercise 6.46 $\forall F, G(S(F, G) \rightarrow F \prec G)$.

Exercise 6.47 $\forall F, G((F \not\approx F \wedge F \approx G) \rightarrow G \not\approx G)$.

Exercise 6.48 $\forall F, G((\mathfrak{F}_{\prec}(F) \wedge F \approx G) \rightarrow \mathfrak{F}_{\prec}(G))$.

We will need to adopt another axiom. First, though, we consider another interpretation: the INTENDED one. Our objects are the natural numbers and ∞. (You can let ∞ be anything that is not a natural number.) Our sets are all the sets of these objects. $\in$ is membership. # assigns objects to sets as follows.

$$\#F = \begin{cases} n & \text{if } F \text{ has } n \text{ members} \\ \infty & \text{if } F \text{ is infinite} \end{cases}$$

This means that $\approx$ is cardinal equivalence. So $F \prec G$ if and only if F is the same size as a proper subset of G. There are two ways for this to happen: either F is finite and smaller than G or F and G are both infinite. Now consider a comprehension axiom of the form

$$\forall G_1, ..., G_m \forall y_1, ..., y_n \exists F \forall x (x \in F \leftrightarrow \phi).$$

For each choice of sets $G_1, ..., G_m$ and objects $y_1, ..., y_n$, our interpretation provides a set whose members are exactly the objects satisfying the predicate ϕ. Our interpretation does so because it includes *every* set of our objects. So every instance of comprehension is true in our interpretation. Axiom 6.1 is true because our object variables range over all the members of our sets and, furthermore, our interpretation treats $\in$ as membership. So, if our interpretation makes

$$\forall x (x \in F \leftrightarrow x \in G)$$

true, then F and G really do have the same members and, so, are the same.

Exercise 6.49 *Confirm that Axiom 6.2 is true in the intended interpretation.*

After a useful definition, we will introduce a new axiom.

Definition 6.16 $\forall F, G (F \precapprox G \leftrightarrow (F \prec G \vee F \approx G))$.

Axiom 6.3 $\forall F, G \forall x (F \prec (G \cup \{x\}) \rightarrow F \precapprox G)$.

Exercise 6.50 *Confirm that Axiom 6.3 is true in our first (tiny) interpretation.*

Exercise 6.51 *Confirm that Axiom 6.3 is true in the intended interpretation.*

Exercise 6.52 $\forall F ((F \cup \{x\}) \prec (F \cup \{x\}) \rightarrow F \prec F)$.

Theorem 6.19 $\forall F (\mathfrak{F}(F) \rightarrow F \nprec F)$.

Proof Apply Exercises 6.40, 6.47, 6.52, and Theorem 6.18.

Exercise 6.53 $\forall F, G ((\mathfrak{F}(F) \wedge F \precapprox G \wedge G \precapprox F) \rightarrow F \approx G)$.

Exercise 6.54 $\forall F, G (S(F, G) \rightarrow \forall H (H \precapprox F \leftrightarrow H \prec G))$.

Exercise 6.55 $\forall F, G, H ((\mathfrak{F}(F) \wedge S(F, G) \wedge S(H, G)) \rightarrow F \approx H)$.

Exercise 6.56 $\forall x \in \omega \forall y, z ((\sigma x y \wedge \sigma z y) \rightarrow x = z)$.

Exercise 6.57 *If $\mathfrak{F}(F)$, $x \notin F$, and $y \notin G$, then*

$$(F \cup \{x\}) \approx (G \cup \{y\}) \rightarrow F \approx G.$$

Exercise 6.58 $\forall G (\mathfrak{F}_{\prec}(G) \rightarrow \forall x \, \mathfrak{F}_{\prec}(G \cup \{x\}))$.

Theorem 6.20 $\forall F, G ((\mathfrak{F}(G) \wedge F \prec G) \rightarrow \mathfrak{F}(F))$.

Proof By Exercises 6.42, 6.48, 6.58, and Theorem 6.18, $\forall G(\mathfrak{F}(G) \rightarrow \mathfrak{F}_{\prec}(G))$.

Exercise 6.59 *If $\mathfrak{F}(F)$, $x \in F$, and $y \in G$, then*

$$F \approx G \rightarrow (F \setminus \{x\}) \approx (G \setminus \{y\}).$$

Exercise 6.60 $\forall F(\mathfrak{F}(F) \rightarrow \forall x \notin F \forall y \in F \; F \approx ((F \cup \{x\}) \setminus \{y\}))$.

Exercise 6.61 $\forall F(\mathfrak{F}(F) \rightarrow \forall x, y \notin F \; (F \cup \{x\}) \approx (F \cup \{y\}))$.

Exercise 6.62 $\forall F, G, H((\mathfrak{F}(F) \wedge S(F, G) \wedge S(F, H)) \rightarrow G \approx H)$.

Exercise 6.63 *Show that 0 has a unique successor.*

Exercise 6.64 *Show that 0's successor has a unique successor.*

Here is our final axiom.

Axiom 6.4 $\forall F, G \forall x(F \prec G \rightarrow (F \cup \{x\}) \precapprox G)$.

Exercise 6.65 *Confirm that Axiom 6.4 is true in our first (tiny) and second (intended) interpretations.*

Here is a table showing how our axioms behave in our two interpretations (writing 'C' for the infinitely many comprehension axioms).

	Tiny	Intended
C	True	True
6.1	True	True
6.2	False	True
6.3	True	True
6.4	True	True

Since Axiom 6.2 can be false while the other axioms are true, 6.2 does not follow from those axioms. Furthermore, since those other axioms are all true in a universe with just one object and two sets (a universe too small and too simple to harbor unseen absurdities), we can have no serious doubts about their consistency. Note too: if we are confident that the story of the intended interpretation is coherent, we can be confident that all our axioms together are consistent.

Exercise 6.66 $\forall F(\mathfrak{F}(F) \rightarrow \forall G(F \prec G \vee G \prec F \vee F \approx G))$.

Exercise 6.67 $\forall F, G, H((\mathfrak{F}(F) \wedge S(F, G) \wedge F \approx H) \rightarrow S(H, G))$.

Exercise 6.68 $\forall x \in \omega \forall y, z((\sigma xy \wedge \sigma xz) \rightarrow y = z)$.

In your solution of Exercise 6.19, you supplied an FA proof that every natural number has a successor. If you go back and look, you can confirm that all the FA results you used in your proof have already been or can easily be reproduced here in MFA.[13] So we state the following theorem without proof.

---

[13] The following results are sufficient: Theorems 6.1–6.5, 6.8, 6.9, 6.11 and Exercises 6.7, 6.9, 6.12, 6.13, 6.15.

Theorem 6.21 $\forall x \in \omega \exists y \, \sigma xy$.

Just as in **FA**, we can now obtain **MFA** translations of the **PA** axioms S1 and S2. Axioms A1, A2, M1, and M2 (governing addition and multiplication) present new challenges. In **FA**, we used relation variables to define addition and multiplication. Having jettisoned the relational machinery, we need a new approach. Here is one way of defining the relation "z is the sum of x and y" without using relation variables.

Definition 6.17 $\forall x, y, z(\alpha(x, y, z) \leftrightarrow \exists F, G(x = \#F \wedge y = \#G \wedge z = \#(F \cup G) \wedge (F \cap G) = \emptyset))$.

The sum of x and y is the number of objects in the union of two disjoint sets, one with x members and one with y members. The relation α lets us define the relation "x is a multiple of y". Actually, we define the set of multiples of y.

Definition 6.18 $\forall x, y(x \in \omega_{xy} \leftrightarrow \forall F((0 \in F \wedge \forall w \in F \forall z(\alpha(w, y, z) \rightarrow z \in F)) \rightarrow x \in F))$.

The idea is that multiples of y are what tireless immortal beings encounter when they start with 0 and add y to everything they encounter. The multiples of y are the things that appear in every set that has 0 as a member and is closed under the operation "plus-y".

The following definition will be useful. Here the "less than" relation $<$ is defined just as in **FA**.

Definition 6.19 $\forall x, y(x \in \omega_{<y} \leftrightarrow (x \in \omega \wedge x < y))$.

We now notice that if $z \neq 0$, then z is the product of x and y if and only if z is a multiple of y and there are x multiples of y less than z. The product of 3 and 1 is 3 because 3 is a multiple of 1 and there are 3 multiples of 1 less than 3 (namely, 0, 1, and 2). The product of 3 and 2 is 6 because 6 is a multiple of 2 and there are 3 multiples of 2 less than 6 (namely, 0, 2, and 4). On the other hand, this approach does not work so well when $z = 0$. We do not want to say: 0 is the product of x and y if and only if 0 is a multiple of y and there are x multiples of y less than 0. This would imply that 0 is the product of x and y only when $x = 0$ (since there cannot be multiples of y less than 0). Note that the following definition of the relation "z is the product of x and y" features a clause that allows 0 to be the product of x and y even when $x \neq 0$.[14]

Definition 6.20 $\forall x, y, z(\pi(x, y, z) \leftrightarrow (z \in \omega_{xy} \wedge (y = 0 \vee x = \#(\omega_{xy} \cap \omega_{<z}))))$.

If you find this topic interesting and want a big project, you might try to prove that suitable translations of A1, A2, M1, and M2 are theorems of **MFA**. You will want to show that pairs of natural numbers always have a unique natural number sum and a unique natural number product. Here are some exercises to get you started.

---

[14] Cf. the definition of multiplication in §5 of Visser [12].

Exercise 6.69 $\forall F, G((\mathfrak{F}(F) \wedge \mathfrak{F}(G)) \to \mathfrak{F}(F \cup G))$.

Exercise 6.70 *If $\mathfrak{F}(F)$, $x \notin F$, and $y \notin G$, then*

$$F \approx G \to (F \cup \{x\}) \approx (G \cup \{y\}).$$

Exercise 6.71 *If $\mathfrak{F}(G_1)$, $(F_1 \cap G_1) = (F_2 \cap G_2) = \emptyset$, and $G_1 \approx G_2$, then*

$$F_1 \approx F_2 \leftrightarrow (F_1 \cup G_1) \approx (F_2 \cup G_2).$$

Exercise 6.72 $\forall y \in \omega \forall x, z_1, z_2((\alpha(x, y, z_1) \wedge \alpha(x, y, z_2)) \to z_1 = z_2)$.

Exercise 6.73

$$\forall x, y_1 \in \omega \forall z_1, z_2((\alpha(x, y_1, z_1) \wedge \sigma z_1 z_2) \to \exists y_2(\alpha(x, y_2, z_2) \wedge \sigma y_1 y_2)).$$

Exercise 6.74

$$\forall x, y_1 \in \omega \forall y_2, z_1, z_2((\alpha(x, y_1, z_1) \wedge \sigma y_1 y_2 \wedge \sigma z_1 z_2) \to \alpha(x, y_2, z_2)).$$

Exercise 6.75 $\forall x \in \omega \exists F \ x = \#F$.

Exercise 6.76 $\forall x \in \omega \ \alpha(x, 0, x)$.

Exercise 6.77 $\forall x, y \in \omega \exists z \in \omega \ \alpha(x, y, z)$.

6.8 Solutions of Odd-Numbered Exercises

6.1 Properties that apply to the same objects apply to the same number of objects. Properties that apply to the same number of objects apply to the same objects. Given any property F, there is a property that does not apply to the same objects as F, but does apply to the same number of objects. There is a property F that interacts with every property G as follows: if F applies to the same number of objects as G, then G applies to no objects at all. Given any relation R, if something does R to everything, then everything has R done to it by something.

6.3 You might define identity as follows: $\forall x \forall y (x = y \leftrightarrow \forall F(Fx \leftrightarrow Fy))$. Note that $\forall F(Fx \leftrightarrow Fx)$. Furthermore, our definition *says* that identical things have the same properties. If you want to confirm that identical things satisfy the same formulas, suppose $\phi(x)$ and $x = y$. Comprehension gives us an F such that $\forall z(Fz \leftrightarrow \phi(z))$. By our definition, Fx if and only if Fy. So $\phi(x)$ if and only if $\phi(y)$. Just for fun, you might confirm that the following definition would work: $\forall x \forall y (x = y \leftrightarrow \forall F(Fx \to Fy))$.

6.5 Comprehension lets us suppose that $\forall x(\mathfrak{D}x \leftrightarrow (x = \#\mathfrak{N} \vee x = \#\mathfrak{U}))$. As in the previous exercise, $\#\mathfrak{N} \neq \#\mathfrak{D}$. Suppose $\mathfrak{U} \approx \mathfrak{D}$. Then, for some relation

R, $\forall y(\mathfrak{D}y \rightarrow \exists_1 x(Rxy \wedge \mathfrak{U}x))$. Note that $\mathfrak{D}\#\mathfrak{N}$ and $\mathfrak{D}\#\mathfrak{U}$. So $\exists_1 x(Rx\#\mathfrak{N} \wedge \mathfrak{U}x)$ and $\exists_1 x(Rx\#\mathfrak{U} \wedge \mathfrak{U}x)$. Since $\#\mathfrak{N}$ is the only object that has property $\mathfrak{U}$, we infer that $R\#\mathfrak{N}\#\mathfrak{N}$ and $R\#\mathfrak{N}\#\mathfrak{U}$. But $\forall x(\mathfrak{U}x \rightarrow \exists_1 y(Rxy \wedge \mathfrak{D}y))$. In particular, $\exists_1 y(R\#\mathfrak{N}y \wedge \mathfrak{D}y)$. Since such a y is unique, we infer that $\#\mathfrak{N} = \#\mathfrak{U}$—contrary to the previous exercise. We conclude that $\mathfrak{U} \not\approx \mathfrak{D}$. So, by Hume's principle, $\#\mathfrak{U} \neq \#\mathfrak{D}$. We need to do some more work before we can prove that there are infinitely many numbers. You may already be convinced, however, that something like the above reasoning can be repeated indefinitely.

6.7 Suppose $\#F = \#\mathfrak{N}$. By Hume's principle, $F \approx \mathfrak{N}$. For some relation R, $\forall x(Fx \rightarrow \exists_1 y(Rxy \wedge \mathfrak{N}y))$. Since nothing has property $\mathfrak{N}$, this implies that nothing has property F. On the other hand, suppose $\exists_0 x\, Fx$. Let R be any relation at all. It is vacuously true that $\forall x(Fx \rightarrow \exists_1 y(Rxy \wedge \mathfrak{N}y))$ just as it is vacuously true that $\forall y(\mathfrak{N}y \rightarrow \exists_1 x(Rxy \wedge Fx))$. So $F \approx \mathfrak{N}$ and, hence, by Hume's principle, $\#F = \#\mathfrak{N}$.

6.9 Suppose $\#\mathfrak{N} = \#G$. Then, by Exercise 6.7, nothing has property G. So there is no way that $\exists x(Gx \wedge \forall y(Fy \leftrightarrow (Gy \wedge y \neq x)))$.

6.11 Suppose x is a natural number and σxy. Suppose F is a σ-closed property that applies to 0. Then Fx since x has every σ-closed property that 0 has. So Fy since F is σ-closed. We conclude: y has every σ-closed property that 0 has. That is, $0 \leq y$.

6.13 It is vacuously true that $(0 \neq 0 \rightarrow \exists w\, \sigma w0)$. Suppose σxy. Then $\exists w\, \sigma wy$ and, hence, $(y \neq 0 \rightarrow \exists w\, \sigma wy)$.

6.15 Theorem 6.5 guarantees that 0 has the desired property. As an inductive hypothesis, suppose $\forall z((x \leq z \leq x) \rightarrow x = z)$. Suppose σxy and $y \leq z \leq y$. We need to show that $y = z$. By Theorems 6.3 and 6.4, $x \leq z$. By Theorem 6.10, $-\sigma zy$ and, hence, $x \neq z$. So, by our inductive hypothesis, $z \not\leq x$. By Theorem 6.6, $z \not< y$ and, hence, $y = z$.

6.17 If $y < 0$, then $y \neq 0$ and, by Theorem 6.5, $y = 0$. So $\exists_0 y\, Fy$. Now apply Exercise 6.7.

6.19 Suppose $\sigma x_1 x_2$. By Exercise 6.16, $\forall y(y \leq x_1 \leftrightarrow y < x_2)$. Suppose $\forall y(F_1 y \leftrightarrow y \leq x_1)$. Then $\forall y(F_1 y \leftrightarrow y < x_2)$ and, hence, by Theorem 6.12, $\#F_1 = x_2$. Suppose $\forall y(F_2 y \leftrightarrow y \leq x_2)$. Then, by Theorem 6.11, $\sigma x_2 \#F_2$.

6.21 $R0w$ because $(0 = 0 \wedge w = w)$. Suppose $Ry_1 y_2$ (as an inductive hypothesis) and $\sigma y_1 z_1$. Exercise 6.19 lets us suppose that $\sigma y_2 z_2$. Then $(\sigma y_1 z_1 \wedge (Ry_1 y_2 \wedge \sigma y_2 z_2))$ and, hence, $\exists x_1(\sigma x_1 z_1 \wedge \exists x_2(Rx_1 x_2 \wedge \sigma x_2 z_2))$. So $Rz_1 z_2$.

6.23 If $Q0y_2$, then $y_2 = w$ and, hence, $R0y_2$. As an inductive hypothesis, suppose $\forall x_2(Qy_1 x_2 \rightarrow Ry_1 x_2)$. Suppose $\sigma y_1 z_1$ and $Qz_1 z_2$. Then $\exists x_1(\sigma x_1 z_1 \wedge \exists x_2(Qx_1 x_2 \wedge \sigma x_2 z_2))$. By Theorem 6.2, $\exists x_2(Qy_1 x_2 \wedge \sigma x_2 z_2)$ and, hence, $\exists x_2(Ry_1 x_2 \wedge \sigma x_2 z_2)$.

So Rz_1z_2.

6.25 We could say: R is a times-w relation if and only if, for all natural numbers y_1 and y_2,

$$Ry_1y_2 \leftrightarrow ((y_1 = 0 \land y_2 = 0) \lor \exists x_1(\sigma x_1 y_1 \land \exists x_2(Rx_1x_2 \land y_2 = (x_2 + w)))).$$

6.27 Comprehension lets us suppose that

$$\forall x_1 \forall y_2 (Rx_1y_2 \leftrightarrow \exists x_2(Qx_1x_2 \land y_2 = (x_2 + x_1))).$$

R is to be our times-Sw or times-$(w + 1)$ relation. The idea is that $y_2 = x_1 \cdot (w + 1)$ if and only if $y_2 = (x_1 \cdot w) + x_1$. Since Q is a times-w relation:

$$Qy_1y_2 \leftrightarrow ((y_1 = 0 \land y_2 = 0) \lor \exists x_1(\sigma x_1 y_1 \land \exists x_2(Qx_1x_2 \land y_2 = (x_2 + w)))).$$

We want to show that R is a times-Sw relation. That is,

$$Rx_1y_2 \leftrightarrow ((x_1 = 0 \land y_2 = 0) \lor \exists z_1(\sigma z_1x_1 \land \exists z_2(Rz_1z_2 \land y_2 = (z_2 + Sw))))$$

whenever x_1 and y_2 are natural numbers. To verify the left–right half of the biconditional, we assume that Rx_1y_2. This lets us pick a z_2 such that Qx_1z_2 and $y_2 = (z_2 + x_1)$. Note that

$$(x_1 = 0 \land z_2 = 0) \lor \exists z_1(\sigma z_1x_1 \land \exists y_1(Qz_1y_1 \land z_2 = (y_1 + w))).$$

Suppose $x_1 = 0$. If $z_2 \neq 0$, then $\sigma z_1 0$ for some z_1, contrary to Exercise 6.9. So $z_2 = 0$ and, hence, by Theorem 6.15, $y_2 = 0$. Suppose $x_1 \neq 0$. Then we can pick z_1, y_1 such that σz_1x_1, Qz_1y_1, and $z_2 = (y_1 + w)$. We can show:

$$y_2 = z_2 + Sz_1 = S(z_2 + z_1) = S((y_1 + w) + z_1) = S((y_1 + z_1) + w) = (y_1 + z_1) + Sw.$$

Note that $Rz_1(y_1 + z_1)$ since $(Qz_1y_1 \land (y_1 + z_1) = (y_1 + z_1))$. So

$$(\sigma z_1x_1 \land (Rz_1(y_1 + z_1) \land y_2 = ((y_1 + z_1) + Sw)))$$

and, hence,

$$\exists z_1(\sigma z_1x_1 \land \exists z_2(Rz_1z_2 \land y_2 = (z_2 + Sw))).$$

Now we need to prove the right–left half of the biconditional. We consider two cases. First: $x_1 = 0 = y_2$. Note that $Q00$ and, hence, $\exists x_2(Q0x_2 \land 0 = (x_2 + 0))$. So $R00$. As for the second case, pick z_1, z_2 such that σz_1x_1, Rz_1z_2, and $y_2 = (z_2 + Sw)$. This lets us pick an x_2 such that Qz_1x_2 and $z_2 = (x_2 + z_1)$. Note that $Qx_1(x_2 + w)$ since $(\sigma z_1x_1 \land (Qz_1x_2 \land (x_2 + w) = (x_2 + w)))$. Furthermore,

$$y_2 = ((x_2 + z_1) + Sw) = ((x_2 + w) + Sz_1) = ((x_2 + w) + x_1).$$

So Rx_1y_2 since $(Qx_1(x_2 + w) \wedge y_2 = ((x_2 + w) + x_1))$.

6.29 Suppose R is a times-w relation. Suppose $R0y_2$ and $R0y_3$. Then, by Exercise 6.9, $y_2 = 0 = y_3$. As an inductive hypothesis, suppose

$$\forall y_2 \forall y_3 ((Ry_1y_2 \wedge Ry_1y_3) \rightarrow y_2 = y_3).$$

Suppose RSy_1y_2 and RSy_1y_3. We can pick x_1, x_2 such that $\sigma x_1 Sy_1$, Rx_1x_2, and $y_2 = (x_2 + w)$. By Theorem 6.2, $x_1 = y_1$ (since $\sigma y_1 Sy_1$) and, hence, Ry_1x_2. We can also pick x_3, x_4 such that $\sigma x_3 Sy_1$, Rx_3x_4, and $y_3 = (x_4 + w)$. By Theorem 6.2, $x_3 = y_1$ and, hence, Ry_1x_4. By our inductive hypothesis, $x_2 = x_4$ and, hence, $y_2 = (x_2 + w) = (x_4 + w) = y_3$.

6.31 Note the following
$$\forall F \ \#F = \#F.$$

$$\forall F, G(\#F = \#G \rightarrow \#G = \#F).$$

$$\forall F, G, H((\#F = \#G \wedge \#G = \#H) \rightarrow \#F = \#H).$$

6.33 By Definition 6.11, the following are equivalent

$$\exists F, G(w = \#F \wedge z = \#G \wedge S(F, G))$$

$$\exists F, G \exists x \notin F(w = \#F \wedge z = \#G \wedge (F \cup \{x\}) \approx G)$$

$$\exists F \exists x \notin F(w = \#F \wedge z = \#(F \cup \{x\})).$$

6.35 Suppose $x \in \omega$ and σxy. Suppose F is σ-closed and $0 \in F$. Then $x \in F$ because x is a natural number. So $y \in F$ because F is σ-closed. We conclude that y is a member of a σ-closed set whenever 0 is. That is, y is a natural number. So $y \in \omega$.

6.37 Suppose $\#F \in \omega$. If $x \in F$, we are done because then $F = (F \cup \{x\})$. Suppose $x \notin F$. Then $S(F, (F \cup \{x\}))$ and, hence, by Exercise 6.33, $\sigma \#F \#(F \cup \{x\})$. So, by Exercise 6.35, $\#(F \cup \{x\}) \in \omega$.

6.39 If $F \subset G$, then $F \approx F \subset G$.

6.41 Suppose $F \neq \emptyset$. Then $\emptyset \subset F$. So, by Exercise 6.39, $\emptyset \prec F$.

6.43 By Exercise 6.31, if $F \approx G \approx H_1 \subset H$, then $F \approx H_1 \subset H$.

6.45 Suppose $F \prec G \prec H$. Pick F_1, G_1 such that $F \approx F_1 \subset G$ and $G \approx G_1 \subset H$. By Exercise 6.39 and Axiom 6.2, $F_1 \prec G_1$. Pick F_2 such that $F_1 \approx F_2 \subset G_1$. Then $F \approx F_1 \approx F_2 \subset H$. So $F \prec H$.

6.47 If $F \approx G \prec G \approx F$, then, by Exercise 6.43 and Axiom 6.2, $F \prec F$.

6.49 If F is finite, Axiom 6.2 says: if F is smaller than G and G is the same size as H, then F is smaller than H. If F is infinite, Axiom 6.2 says: if F and G are both infinite and G is the same size as H, then F and H are both infinite.

6.51 If F is finite, Axiom 6.3 says: if F is smaller than $G \cup \{x\}$, then F is either smaller than or the same size as G. If F is infinite, Axiom 6.3 says: if F and $G \cup \{x\}$ are both infinite, then F and G are both infinite.

6.53 If $\mathfrak{F}(F)$ and $F \prec G \prec F$, then, by Exercise 6.45, $F \prec F$, contrary to Theorem 6.19.

6.55 Suppose $S(F, G)$ and $S(H, G)$. By Exercise 6.46, $F \prec G$ and $H \prec G$. By Exercise 6.54, $H \precapprox F$ and $F \precapprox H$. Now apply Exercise 6.53.

6.57 Suppose $(F \cup \{x\}) \approx (G \cup \{y\})$. By Exercise 6.34, $S(F, (G \cup \{y\}))$ since $S(F, (F \cup \{x\}))$. By Exercise 6.55, $F \approx G$ since $S(G, (G \cup \{y\}))$.

6.59 By Exercise 6.39, $(F \backslash \{x\}) \prec F$ since $x \in F$. So, by Theorem 6.20, $\mathfrak{F}(F \backslash \{x\})$ since $\mathfrak{F}(F)$. Note that $((F \backslash \{x\}) \cup \{x\}) = F \approx G = ((G \backslash \{y\}) \cup \{y\})$. So we need only apply Exercise 6.57.

6.61 By Exercise 6.37, $\mathfrak{F}(F \cup \{y\})$. If $x = y$, we are done. Suppose $x \neq y$. Then $x \notin (F \cup \{y\})$. Furthermore, $y \in (F \cup \{y\})$. So, by Exercise 6.60, $(F \cup \{y\}) \approx (((F \cup \{y\}) \cup \{x\}) \backslash \{y\})$. Since $y \notin F$, $(((F \cup \{y\}) \cup \{x\}) \backslash \{y\}) = (F \cup \{x\})$.

6.63 Note that $S(\emptyset, \{0\})$ since $0 \notin \emptyset$ and $(\emptyset \cup \{0\}) \approx \{0\}$. By Exercise 6.33, $\sigma 0 \# \{0\}$. Suppose $\sigma 0 z$. Pick F, G such that $0 = \#F$, $z = \#G$, and $S(F, G)$. By Exercise 6.44, $F = \emptyset$ since $F \approx \emptyset$. By Exercises 6.36 and 6.62, $G \approx \{0\}$ since $S(\emptyset, G)$. So $z = \#G = \#\{0\}$.

6.65 In the tiny interpretation, $(F \cup \{x\}) \precapprox G$ because every set is equivalent to every set. Now consider the intended interpretation. If F is finite, Axiom 6.4 says: if F is smaller than G, then $F \cup \{x\}$ is either smaller than or the same size as G. If F is infinite, Axiom 6.4 says: if F and G are both infinite, then $F \cup \{x\}$ and G are both infinite.

6.67 Suppose $x \notin F$ and $(F \cup \{x\}) \approx G$. By Exercises 6.37 and 6.38, $\mathfrak{F}(G)$. If every object belongs to H, then $F \subset H$ and, hence, by Exercise 6.39 and Axiom 6.2, $F \prec F$—contrary to Theorem 6.19. Suppose $y \notin H$. Suppose $G \prec (H \cup \{y\})$. By Axiom 6.3 and Exercise 6.39, $G \precapprox H \approx F \prec (F \cup \{x\}) \approx G$. So, by Axiom 6.2

and Exercises 6.43 and 6.45, $G \prec G$—contrary to Theorem 6.19. On the other hand, suppose $(H \cup \{y\}) \prec G$. By Axioms 6.2 and 6.3, $(H \cup \{y\}) \precsim F$. By Exercise 6.39, $H \prec (H \cup \{y\})$. So by Axiom 6.2 and Exercises 6.43 and 6.45, $F \prec F$—contrary to Theorem 6.19. Exercise 6.66 offers just one alternative: $(H \cup \{y\}) \approx G$. So $S(H, G)$.

6.69 We are going to apply Theorem 6.18 using the following formula $\phi(H)$:

$$\mathfrak{F}(H) \wedge \forall F, G((\mathfrak{F}(F) \wedge G \approx H) \to \mathfrak{F}(F \cup G)).$$

Exercises 6.36 and 6.44 make it easy to confirm that $\phi(\emptyset)$. Suppose, as a kind of inductive hypothesis, that $\phi(H)$. If $x \in H$, then $\phi(H \cup \{x\})$ because $(H \cup \{x\}) = H$. Suppose $x \notin H$. Then $((H \cup \{x\})\setminus\{x\}) = H$. Suppose $\mathfrak{F}(F)$ and $G \approx (H \cup \{x\})$. By Exercises 6.37 and 6.38, $\mathfrak{F}(G)$. By Exercise 6.44, $G \neq \emptyset$. Suppose $y \in G$. Then, by Exercise 6.59, $(G\setminus\{y\}) \approx H$. So, by our inductive hypothesis, $\mathfrak{F}(F \cup (G\setminus\{y\}))$ and, hence, by Exercise 6.37, $\mathfrak{F}(F \cup G)$. That is, $\phi(H \cup \{x\})$. Now suppose $\phi(H_1)$ and $H_1 \approx H_2$. By Exercise 6.38, $\mathfrak{F}(H_2)$. Suppose $\mathfrak{F}(F)$ and $G \approx H_2$. Then $G \approx H_1$ and, hence, $\mathfrak{F}(F \cup G)$. So $\phi(H_2)$. Theorem 6.18 lets us conclude: $\forall H(\mathfrak{F}(H) \to \phi(H))$. So $\forall H(\mathfrak{F}(H) \to \forall F((\mathfrak{F}(F) \wedge H \approx H) \to \mathfrak{F}(F \cup H)))$. Now apply Exercise 6.31.

6.71 We use induction on x to show: if $x \in \omega$, then G_1 satisfies the theorem whenever $\#G_1 = x$. First, suppose $\#G_1 = 0$. Then, by Definition 6.11 and Exercise 6.44, $G_1 = \emptyset$. If $G_1 \approx G_2$, then, by Exercises 6.31 and 6.44, $G_2 = \emptyset$ and, hence, $F_1 \approx F_2$ if and only if $(F_1 \cup G_1) \approx (F_2 \cup G_2)$. Now, as an inductive hypothesis, suppose x has the desired property. Suppose $\sigma x y$. Use Exercise 6.33 to pick G_1, H_1 such that $x = \#H_1$, $y = \#G_1$, and $S(H_1, G_1)$. Definition 6.12 lets us suppose $w \notin H_1$ and $(H_1 \cup \{w\}) \approx G_1$. By Exercise 6.44, $G_1 \neq \emptyset$. Suppose $z_1 \in G_1$. Then, by Exercise 6.59, $(G_1\setminus\{z_1\}) \approx H_1$. By Definition 6.11, $\#(G_1\setminus\{z_1\}) = x$ and, hence, by our inductive hypothesis, $(G_1\setminus\{z_1\})$ satisfies the theorem. That is, given any F_1, F_2, H_2, if $(F_1 \cap (G_1\setminus\{z_1\})) = (F_2 \cap H_2) = \emptyset$, and $(G_1\setminus\{z_1\}) \approx H_2$, then

$$F_1 \approx F_2 \leftrightarrow (F_1 \cup (G_1\setminus\{z_1\})) \approx (F_2 \cup H_2).$$

Suppose $(F_1 \cap G_1) = (F_2 \cap G_2) = \emptyset$ and $G_1 \approx G_2$. Since $G_1 \neq \emptyset$, Exercise 6.44 guarantees that $G_2 \neq \emptyset$. Suppose $z_2 \in G_2$. Then, by Exercise 6.59, $(G_1\setminus\{z_1\}) \approx (G_2\setminus\{z_2\})$. So

$$F_1 \approx F_2 \leftrightarrow (F_1 \cup (G_1\setminus\{z_1\})) \approx (F_2 \cup (G_2\setminus\{z_2\})).$$

But, by Theorem 6.20 and Exercises 6.39, 6.57, 6.69, and 6.70, $(F_1 \cup (G_1\setminus\{z_1\})) \approx (F_2 \cup (G_2\setminus\{z_2\}))$ if and only if

$$((F_1 \cup (G_1\setminus\{z_1\})) \cup \{z_1\}) \approx ((F_2 \cup (G_2\setminus\{z_2\})) \cup \{z_2\}).$$

So $F_1 \approx F_2$ if and only if $(F_1 \cup G_1) \approx (F_2 \cup G_2)$. That is, G_1 satisfies the theorem.

6.73 We begin with some assumptions.

$$
\begin{array}{ll}
(1)\ x = \#F & (5)\ z_1 = \#H_1 \\
(2)\ y_1 = \#G & (6)\ z_2 = \#H_2 \\
(3)\ z_1 = \#(F \cup G) & (7)\ w_1 \notin H_1 \\
(4)\ (F \cap G) = \emptyset & (8)\ (H_1 \cup \{w_1\}) \approx H_2
\end{array}
$$

Note that, by (1), (2), and Exercise 6.69, $\mathfrak{F}(F \cup G)$ since $x, y_1 \in \omega$. By (3), (5), and Definition 6.11, $(F \cup G) \approx H_1$. If $w_2 \in (F \cup G)$ for every w_2, then, by (7), $H_1 \subset (F \cup G)$ and, hence, by Exercises 6.39 and 6.43, $(F \cup G) \prec (F \cup G)$, contrary to Theorem 6.19. Suppose $w_2 \notin (F \cup G)$. Then, by (2) and Exercise 6.33, $\sigma y_1 \#(G \cup \{w_2\})$. Furthermore, by (7), (8), and Exercise 6.70,

$$(F \cup (G \cup \{w_2\})) = ((F \cup G) \cup \{w_2\}) \approx (H_1 \cup \{w_1\}) \approx H_2.$$

So, by (6) and Definition 6.11, $z_2 = \#(F \cup (G \cup \{w_2\}))$. By (4), $(F \cap (G \cup \{w_2\})) = \emptyset$. So, by (1), $\alpha(x, \#(G \cup \{w_2\}), z_2)$.

6.75 We use induction. Recall that $0 = \#\emptyset$ and then apply Definition 6.3.

6.77 Induction on y. Exercise 6.76 handles the case when $y = 0$. Suppose $\alpha(x, y_1, z_1)$ and $\sigma y_1 y_2$. Theorem 6.21 lets us suppose $\sigma z_1 z_2$. Now apply Exercise 6.74.

References

1. Boolos, G. S., & Heck, R. G. (1998). Die Grundlagen der Arithmetik §§82-83. In M. Schirn (Ed.), *Philosophy of mathematics today* (pp. 407–428). Oxford: Oxford University Press.
2. Burgess, J. P. (2005). *Fixing frege*. Princeton NJ: Princeton University Press.
3. Frege, G. (1884). *Die Grundlagen der Arithmetik*. Breslau: Wilhelm Koebner.
4. Frege, G. (1893). *Die Grundgesetze der Arithmetik* (Vol. 1). Jena: Hermann Pohle.
5. Frege, G. (1903). *Die Grundgesetze der Arithmetik* (Vol. 2). Jena: Hermann Pohle.
6. Frege, G. (1980). *The foundations of arithmetic*. Evanston, IL: Northwestern University Press.
7. George, A., & Velleman, D. J. (2002). *Philosophies of mathematics*. Oxford: Blackwell.
8. Quine, W. V. O. (1955). On Frege's way out. *Mind, 64*, 145–159.
9. Rang, B., & Thomas, W. (1981). Zermelo's discovery of the 'Russell paradox'. *Historia Mathematica, 8*, 15–22.
10. Shapiro, S. (2011). The company kept by cut abstraction (and its relatives). *Philosophia Mathematica, 19*, 107–138.
11. van Heijenoort, J. (Ed.). (1967). *From Frege to Gödel*. Cambridge MA: Harvard University Press.
12. Visser, A. (2011). Hume's principle, beginnings. *Review of Symbolic Logic, 4*, 114–129.
13. Weir, A. (2003). Neo-Fregeanism: An embarrassment of riches. *Notre Dame Journal of Formal Logic, 44*, 13–48.

Chapter 7
Intuitionist Logic

7.1 Inference

We are going to use a fable to show three things. First, speakers of a language free of the usual logical vocabulary (in particular, without the resources for making "if... then" statements) can, nonetheless, have a well-developed conception of correct inference. Second, speakers of such a language can adopt a locution allowing them to assert that one sentence is inferable from another. Third, the result of this innovation will be a fragment of something called "intuitionist logic." The upshot is that intuitionist logic can, at least in part, be construed as a medium for discussing and reasoning about reasoning. This idea of a logic emerging from a pre-existing practice of inference is not new to us: we already saw this happen in Chap. 1. There, however, it was *classical* logic that emerged from PRA—and, as we shall see, classical logic is incompatible with the intuitionist logic we are trying to develop in this chapter. We will have to be careful not to follow the pattern of Chap. 1 too closely.

Suppose we are observing some people, the Peregrins, who reliably distinguish between good and bad inferences (accepting only the former), but lack the vocabulary to *say* that a conclusion follows from some premises.[1] Here is a possible dialogue.

VILLAGE ELDER: The repeating decimal 0.123123123... is rational.
INQUISITIVE YOUTH: How do you know that?
VILLAGE ELDER:
$$0.123123123\ldots = \frac{123}{10^3 - 1}.$$

INQUISITIVE YOUTH: OK, but how do you know *that*?

The youth accepts the inference from premise to conclusion. (It is apparent that 123 divided by $10^3 - 1$ is rational.) Nonetheless, the youth would like a reason to believe

---

[1] The name 'Peregrin' is a tribute to JAROSLAV PEREGRIN whose paper [7] outlines the approach to intuitionist logic we pursue in this chapter.

S. Pollard, *A Mathematical Prelude to the Philosophy of Mathematics*,
DOI: 10.1007/978-3-319-05816-0_7,
© Springer International Publishing Switzerland 2014

the premise itself. (It is not so apparent that the repeating decimal is exactly that rational number.) The elder is only too happy to oblige.

VILLAGE ELDER:

$$(10^3 - 1) \times 0.123123123\ldots = 123.123123\ldots - 0.123123123\ldots = 123.$$

INQUISITIVE YOUTH: Right.

The youth accepts the inference from the new premise to the prior premise. (Just divide through by $10^3 - 1$ in the new premise to obtain the prior one.) The youth also accepts the new premise itself. (To verify the new premise, you only need to perform the indicated operations.)

As I already mentioned, it is a distinctive feature of Peregrinese that the inquisitive youth can *endorse* the inference from the new premise to the prior one, but cannot *assert* that the one follows from the other. We can do so easily enough

$$0.123123123\ldots = \frac{123}{10^3 - 1} \quad \text{if} \quad (10^3 - 1) \times 0.123123123\ldots = 123$$

but there is no Peregrinese translation of our 'if'. We can also assert that the elder's initial statement follows from the earlier premise:

$$0.123123123\ldots \text{ is rational if it is equal to } \frac{123}{10^3 - 1}.$$

Again, the youth cannot say this.

If the inference from some Peregrinese sentences $\phi_1, \ldots, \phi_n$ to the Peregrinese sentence ψ is judged correct by the Peregrins, *we* record this by writing

$$\{\phi_1, \ldots, \phi_n\} \vdash \psi.$$

Statements of this form are known as SEQUENTS. In some cases, the Peregrins will assent to a sentence ψ without benefit of inferential support. We then record the sequent

$$\emptyset \vdash \psi.$$

By treating inference as a relation between a *set* of premises and a conclusion, we capture two important features of Peregrinese behavior. First, in assessing inferences, the Peregrins are indifferent to the order in which premises are given. If they accept the inference from premises ϕ_1, ϕ_2, presented in that order, to conclusion ψ, they will accept the inference from premises ϕ_2, ϕ_1, presented in that order, to conclusion ψ. This is reflected in the set theoretic fact that $\{\phi_1, \phi_2\} = \{\phi_2, \phi_1\}$. Second, in assessing inferences, the Peregrins are indifferent to repetitions of premises. They will accept the inference from premises ϕ_1, ϕ_1, ϕ_2 to conclusion ψ if and only if they will accept the inference from ϕ_1, ϕ_2 to ψ. This is reflected in the set theoretic fact that $\{\phi_1, \phi_1, \phi_2\} = \{\phi_1, \phi_2\}$.

Three other features of Peregrinese behavior are worth mentioning. We will express them as axioms governing the inferability relation $\vdash$. *First*, they will accept any inference from a sentence to *itself*.

Axiom 7.1 (TAUT)

$$\{\phi\} \vdash \phi.$$

'Taut' is short for 'tautology'. *Second*, if they accept an inference, they will continue to accept it when premises are added.

Axiom 7.2 (THIN)

$$\Phi \vdash \psi \quad \Rightarrow \quad \Phi \cup \{\phi\} \vdash \psi.$$

We use capital Greek letters (such as 'Φ') to stand for *sets* of sentences. 'Thin' is short for 'thinning'. *Third*, if they accept the inference from $\phi_1, \ldots, \phi_n$ to ψ, they are willing to replace ψ with $\phi_1, \ldots, \phi_n$ in any correct inference where ψ serves as a premise.

Axiom 7.3 (CUT)

$$\Psi \vdash \psi, \quad \Phi \cup \{\psi\} \vdash \chi \quad \Rightarrow \quad \Phi \cup \Psi \vdash \chi.$$

A particularly simple form of the Cut principle is:

$$\{\phi\} \vdash \psi, \quad \{\psi\} \vdash \chi \quad \Rightarrow \quad \{\phi\} \vdash \chi.$$

For example, having accepted the inferences

$$(10^3 - 1) \times 0.123123123\ldots = 123 \vdash 0.123123123\ldots = \frac{123}{10^3 - 1}$$

$$0.123123123\ldots = \frac{123}{10^3 - 1} \vdash 0.123123123\ldots \text{ is rational}$$

the inquisitive youth will accept the inference

$$(10^3 - 1) \times 0.123123123\ldots = 123 \vdash 0.123123123\ldots \text{ is rational.}$$

The Peregrins treat the inferability relation as transitive.

7.2 Conjunctions

The Peregrins are careful to distinguish between sentences and clauses within sentences, giving a characteristic wink to indicate the end of a sentence. It is always clear when

Sentence. Sentence.

is intended rather than

Clause; clause.

We indicate the former structure by writing

$$\phi_1, \phi_2$$

and the latter by writing

$$(\phi_1 \,\&\, \phi_2).$$

When they assess inferences, the Peregrins treat

Premise. Premise.

as equivalent to

Premise; premise.

We capture this practice in the following axiom.

Axiom 7.4 (AND)

$$\{\phi_1, \phi_2\} \vdash \psi \quad \Longleftrightarrow \quad \{(\phi_1 \,\&\, \phi_2)\} \vdash \psi.$$

The Peregrins are willing to accept the inference from ϕ_1, ϕ_2 to ψ if and only if they are willing to accept the inference from $(\phi_1 \,\&\, \phi_2)$ to ψ.

Theorem 7.1 $\{(\phi_1 \,\&\, \phi_2)\} \vdash \phi_1$.

Proof Here is a formal derivation.

$$
\begin{array}{lll}
1. & \{\phi_1\} \vdash \phi_1 & \text{Taut} \\
2. & \{\phi_1, \phi_2\} \vdash \phi_1 & \text{1 Thin} \\
3. & \{(\phi_1 \,\&\, \phi_2)\} \vdash \phi_1 & \text{2 And}
\end{array}
$$

This will be our standard format for verifying sequents: a series of numbered lines, each featuring a sequent and a justification. The justification '1 Thin' on line 2 means that we obtained the sequent on line 2 by applying the principle Thin to the sequent on line 1.

Theorem 7.2 $\{(\phi_1 \,\&\, \phi_2)\} \vdash \phi_2$.

Exercise 7.1 $\{\phi_1, \phi_2\} \vdash (\phi_1 \,\&\, \phi_2)$.

Theorem 7.3 $\Phi \cup \{\phi_1, \phi_2\} \vdash \psi \;\Rightarrow\; \Phi \cup \{(\phi_1 \,\&\, \phi_2)\} \vdash \psi$.

Proof In this case, we are not verifying a sequent: we are providing a recipe for passing from one sequent to another in the course of such a verification. When we write down the starting sequent, our justification is 'Prem' (for 'premise').

$$
\begin{array}{lll}
1. & \Phi \cup \{\phi_1, \phi_2\} \vdash \psi & \text{Prem} \\
2. & \{(\phi_1 \,\&\, \phi_2)\} \vdash \phi_1 & \text{Thm 7.1} \\
3. & \Phi \cup \{(\phi_1 \,\&\, \phi_2), \phi_2\} \vdash \psi & \text{1,2 Cut} \\
4. & \{\phi_1 \,\&\, \phi_2\} \vdash \phi_2 & \text{Thm 7.2} \\
5. & \Phi \cup \{(\phi_1 \,\&\, \phi_2)\} \vdash \psi & \text{3,4 Cut}
\end{array}
$$

We allow ourselves to write down sequents already verified, citing the relevant theorem ('Thm') or exercise ('Ex'). Here is a more readable version of the derivation without set theoretic notation and Greek letters.

$$
\begin{array}{lll}
1. & P, Q, R \vdash S & \text{Prem} \\
2. & (Q \,\&\, R) \vdash Q & \text{Thm 7.1} \\
3. & P, (Q \,\&\, R), R \vdash S & \text{1,2 Cut} \\
4. & (Q \,\&\, R) \vdash R & \text{Thm 7.2} \\
5. & P, (Q \,\&\, R) \vdash S & \text{3,4 Cut}
\end{array}
$$

We will use this streamlined format whenever it seems helpful.

We have shown

$$P, Q, R \vdash S \quad \Rightarrow \quad P, (Q \,\&\, R) \vdash S$$

without using any special information about 'P', 'Q', 'R', and 'S'. The proof would have worked just as well with other formulas. So any time you have a sequent of form

$$P, Q, R \vdash S$$

with 'P', 'Q', 'R', and 'S' as shown or replaced by other formulas (or even with the idle letter 'P' replaced by several formulas), you can derive the corresponding sequent

$$P, (Q \,\&\, R) \vdash S$$

citing 'Thm 7.3'. For example, from

$$S_1, \ldots, S_n, (Q \,\&\, P), R \vdash P$$

you can derive

$$S_1, \ldots, S_n, ((Q \,\&\, P) \,\&\, R) \vdash P$$

on the authority of Theorem 7.3.

Exercise 7.2 $P, (Q \,\&\, R) \vdash S \Rightarrow P, Q, R \vdash S$.

Exercise 7.3 $S_1, P \vdash Q$ and $S_2, P \vdash R \Rightarrow S_1, S_2, P \vdash (Q \& R)$.

Exercise 7.4 $(P \& Q) \vdash (Q \& P)$.

Exercise 7.5 $(P \& (Q \& R)) \vdash ((P \& Q) \& R)$.

7.3 Conditionals

We now start to wonder what sort of expansion of Peregrinese would allow the Peregrins to assert that one sentence is inferable from others (that is, assert it by using a sentence that *says* it, not just by assenting to the inference when it is made). One device would be a new Peregrinese symbol '$\triangleright$' that expresses the relation of inferability. Affirming $(\phi \triangleright \psi)$ would then be equivalent to endorsing the inference from ϕ to ψ. So '$\triangleright$' should have the following properties.

First, Peregrins should accept $(\phi \triangleright \psi)$ whenever they accept the inference from ϕ to ψ.

Axiom 7.5 (DED)[2]

$$\{\phi\} \vdash \psi \quad \Rightarrow \quad \emptyset \vdash (\phi \triangleright \psi).$$

Second, from ϕ and $(\phi \triangleright \psi)$, Peregrins should be prepared to infer ψ.

Axiom 7.6 (MP)[3]

$$\{\phi, (\phi \triangleright \psi)\} \vdash \psi.$$

Third, if Peregrins accept the inference from ϕ_0 to ψ_0 whenever they accept the inference from ϕ_1 to ψ_1, then they should accept the inference from $(\phi_1 \triangleright \psi_1)$ to $(\phi_0 \triangleright \psi_0)$. More generally, given

$$\{\phi_1\} \vdash \psi_1, \ldots, \{\phi_n\} \vdash \psi_n \quad \Rightarrow \quad \{\phi_0\} \vdash \psi_0$$

---

[2] The name 'Ded' recalls a result, known as the *deduction* theorem, of which our axiom is a special case. We still need to confirm the full deduction theorem: that is, we still need to confirm that $(\phi \triangleright \psi)$ is inferable when the derivation of ψ from ϕ depends on additional premises Φ:

$$\Phi \cup \{\phi\} \vdash \psi \quad \Rightarrow \quad \Phi \vdash (\phi \triangleright \psi).$$

Axioms 7.7, 7.8, and 7.9 will help us do so. This will show that the Peregrins should accept the inference from Φ to $(\phi \triangleright \psi)$ whenever they accept the inference from $\Phi \cup \{\phi\}$ to ψ. Perhaps we should have just adopted *that* as an axiom. I saw two reasons not to take that course. First, I thought Axiom 7.9 deserved separate discussion because of its special role in distinguishing between inferability and strict implication (to be discussed below). Second, I thought it quite evident that Axioms 7.7 and 7.8 belong in a logic of inferability, but less immediately evident that the deduction theorem does. Others may see the matter differently.

[3] 'MP' stands for '*modus ponens*'.

we should have

$$\{(\phi_1 \rhd \psi_1), \ldots, (\phi_n \rhd \psi_n)\} \quad \vdash \quad (\phi_0 \rhd \psi_0).$$

This means that Cut and Exercise 7.3 yield the following axioms.

Axiom 7.7 ($\rhd$CUT)

$$\{(\phi \rhd \psi), (\psi \rhd \chi)\} \vdash (\phi \rhd \chi).$$

Axiom 7.8 ($\rhd$AND)

$$\{(\chi \rhd \phi), (\chi \rhd \psi)\} \vdash (\chi \rhd (\phi \,\&\, \psi)).$$

As I already mentioned, our target is something known as INTUITIONIST LOGIC. At the moment, we are trying to introduce the rules for the intuitionist conditional. We are still a bit short of our goal. So far, our connective $\rhd$ is indistinguishable from connectives you may have encountered in a logic class: the strict implications $\multimap$ of various modal logics. We need to add an axiom to make it clear that $\rhd$ is not $\multimap$ (because we know that the intuitionist conditional is not $\multimap$).

Axiom 7.9 (INT)

$$\{\psi\} \vdash (\phi \rhd \psi).$$

You might think that Int is unnecessary since Ded and Thin yield:

$$\emptyset \vdash \psi \quad \Rightarrow \quad \emptyset \vdash (\phi \rhd \psi).$$

If Peregrins accept ψ, they accept $(\phi \rhd \psi)$. It does not follow, however, that they should accept the inference from ψ to $(\phi \rhd \psi)$. Students of modal logics such as T, S4, and S5 can confirm that it does not follow. If ψ is logically true, then so is $(\phi \multimap \psi)$. However, $(\phi \multimap \psi)$ will follow from ψ only in special cases. (The truth of ψ does not guarantee that ψ will be true in every possible world where ϕ is true.)

Is Int faithful to the Peregrinese practice of inference? Ded does not guarantee this. Perhaps we could acquire *behavioral* evidence that Int captures some aspect of Peregrinese reasoning. Here is a dialogue that might be relevant.

VILLAGE ELDER: ϕ.
INQUISITIVE YOUTH: Well, I'm willing to suppose so, just for the sake of argument. So what?
VILLAGE ELDER: ψ!
INQUISITIVE YOUTH: Huh?

A plausible interpretation of this discourse is that the elder has inferred ψ from ϕ and that the youth is questioning that inference. The elder now offers support for that inference. (Suppose "Huh?" is a standard way of soliciting support for an inference. We would then expect the elder to respond by offering such support.)

VILLAGE ELDER: ψ.

INQUISITIVE YOUTH: OK, though I'd be interested to know why I should believe ψ.

The youth accepts something. It is not ψ. The youth could be acknowledging that ψ provides a good reason to accept the inference from ϕ to ψ. If so, Int would be justified by the way Peregrins assess inferences. We will take it for granted that Int is justified in this way.

Here is a result we will need for the proof of Theorem 7.4.

Exercise 7.6 $\vdash (Q \rhd Q)$.

As part of our streamlining process, we leave a blank to the left of '$\vdash$' when there should officially be '$\emptyset$'. An official version of the sequent in the above exercise would be '$\emptyset \vdash (\psi \rhd \psi)$'.

Theorem 7.4 $P \vdash (Q \rhd (P \& Q))$.

Proof Here is a formal derivation.

$$
\begin{array}{lll}
1. & (Q \rhd P), (Q \rhd Q) \vdash (Q \rhd (P \& Q)) & \rhd\text{And} \\
2. & \vdash (Q \rhd Q) & \text{Ex 7.6} \\
3. & (Q \rhd P) \vdash (Q \rhd (P \& Q)) & 1,2 \text{ Cut} \\
4. & P \vdash (Q \rhd P) & \text{Int} \\
5. & P \vdash (Q \rhd (P \& Q)) & 3,4 \text{ Cut}
\end{array}
$$

Theorem 7.5 $P, Q \vdash R \Rightarrow P \vdash (Q \rhd R)$.

Proof The formal derivation below provides a recipe for getting from one sequent to the other.

$$
\begin{array}{lll}
1. & P, Q \vdash R & \text{Prem} \\
2. & (P \& Q) \vdash R & 1 \text{ And} \\
3. & \vdash ((P \& Q) \rhd R) & 2 \text{ Ded} \\
4. & (Q \rhd (P \& Q)), ((P \& Q) \rhd R) \vdash (Q \rhd R) & \rhd\text{Cut} \\
5. & (Q \rhd (P \& Q)) \vdash (Q \rhd R) & 3,4 \text{ Cut} \\
6. & P \vdash (Q \rhd (P \& Q)) & \text{Thm 7.4} \\
7. & P \vdash (Q \rhd R) & 5,6 \text{ Cut}
\end{array}
$$

Theorem 7.6 $P_1, \ldots, P_n, Q \vdash R \Rightarrow P_1, \ldots, P_n \vdash (Q \rhd R)$.

Proof The proof is by induction on the number of premises $P_1, \ldots, P_n$. Theorem 7.5 handles the case when $n = 1$. Suppose

$$P_1, \ldots, P_{n-1}, P_n, P_{n+1}, Q \vdash R.$$

Then, by Theorem 7.3,

$$P_1, \ldots, P_{n-1}, (P_n \,\&\, P_{n+1}), Q \vdash R.$$

Notice that there are now only n premises other than Q. So, by inductive hypothesis,

$$P_1, \ldots, P_{n-1}, (P_n \,\&\, P_{n+1}) \vdash (Q \rhd R).$$

Now just apply Exercise 7.2. (Note that in Exercise 7.2 the one sentence P could have been a set of sentences Φ.)

Exercise 7.7 $P_1, \ldots, P_n \vdash (Q \rhd R) \;\Rightarrow\; P_1, \ldots, P_n, Q \vdash R.$

Theorem 7.7 $P_1, \ldots, P_n \vdash (Q \rhd R) \;\Longleftrightarrow\; P_1, \ldots, P_n, Q \vdash R.$

Exercise 7.8 $(P \rhd (Q \rhd R)) \vdash ((P \rhd Q) \rhd (P \rhd R)).$

7.4 Negations

No one knows when the Peregrins first realized that some statements of arithmetic allowed them to infer *every* statement of arithmetic.

VILLAGE ELDER: $0 = 1$.
INQUISITIVE YOUTH: [*incredulously*] What?
VILLAGE ELDER: [*patiently*] Just suppose $0 = 1$.
INQUISITIVE YOUTH: OK, I'll play along, just for the sake of argument.
VILLAGE ELDER: $3 + 0 = 3 + 1$.
INQUISITIVE YOUTH: Yep.
VILLAGE ELDER: $3 = 4$.
INQUISITIVE YOUTH: Yep.
VILLAGE ELDER: $4 = 2 \times 2$.
INQUISITIVE YOUTH: Yep.
VILLAGE ELDER: $3 = 2 \times 2$.
INQUISITIVE YOUTH: Yep.
VILLAGE ELDER: 3 is even.
INQUISITIVE YOUTH: Yep.
VILLAGE ELDER: $3 + 0 = 4 + 1$.
INQUISITIVE YOUTH: Yep.
VILLAGE ELDER: $3 = 5$.
INQUISITIVE YOUTH: Yep.
VILLAGE ELDER: 5 is even.
INQUISITIVE YOUTH: Yeah, yeah, I think I get the point.

We have
$$0 = 1 \vdash \psi$$

whenever ψ is a statement of Peregrinese arithmetic. So, by Cut,

$$\Phi \vdash 0 = 1 \quad \Rightarrow \quad \Phi \vdash \psi$$

whenever ψ is a statement of Peregrinese arithmetic. For Peregrins, inferring '$0 = 1$' from an interlocutor's premises is like saying, "If you believe that, you'll believe anything about the natural numbers." At some point, they acquired the ability to say, in effect, "If you believe that, you'll believe *anything*." They have a pronouncement

Balderdash!

or, more briefly,

$$\bot \qquad \qquad .$$

from which they are prepared to infer any declarative Peregrinese sentence.

Axiom 7.10 (EFQ)[4]
$$\bot \vdash \psi.$$

Theorem 7.8 $\Phi \vdash \bot \Rightarrow \Phi \vdash \psi.$

After reading our discussion of conditionals, the Peregrins quickly adopted the neologism '$\triangleright$' and soon combined it with '$\bot$' to expand their logical vocabulary even more. They began to use

$$\oslash \phi$$

as shorthand for

$$(\phi \triangleright \bot).$$

Definition 7.1 You can freely interchange $\oslash \phi$ and $(\phi \triangleright \bot)$.

$\oslash$ is a kind of stop sign. A statement of the form

$$\oslash \phi$$

is a warning that, if you accept ϕ, you risk trivializing future inferences by allowing yourself to infer everything. A statement of the form

$$(\phi \triangleright \oslash \psi)$$

is a warning that, if you accept ϕ, you had better not accept ψ.

---

[4] 'EFQ' stands for '*ex falso quodlibet*': "from a falsehood anything [follows]."

Theorem 7.9 $\vdash \oslash \perp$.

Proof '$\oslash \perp$' is shorthand for '$(\perp \triangleright \perp)$'. So we just apply Exercise 7.6.

Notice that I did not use EFQ in the above proof. I would like you, too, to avoid EFQ in the following exercises. You should feel free, however, to use Definition 7.1 to replace sentences of the form $\oslash \phi$ with ones of the form $(\phi \triangleright \perp)$.

Exercise 7.9 $(P \triangleright Q) \vdash (\oslash Q \triangleright \oslash P)$.

Exercise 7.10 $R_1, \ldots, R_n, P \vdash Q \;\Rightarrow\; R_1, \ldots, R_n, \oslash Q \vdash \oslash P$.

Exercise 7.11 $(P \triangleright \oslash Q) \vdash (Q \triangleright \oslash P)$.

Exercise 7.12 $(P \triangleright \oslash P) \vdash \oslash P$. (*Remember Exercise 7.8.*)

Exercise 7.13 $\perp \vdash \oslash P$.

Exercise 7.14 $P \vdash \oslash \oslash P$.

Exercise 7.15 $\oslash \oslash \oslash P \vdash \oslash P$.

Exercise 7.16 $\oslash \oslash \perp \vdash \perp$.

Exercise 7.17 $P_1, \ldots, P_n \vdash \oslash Q \;\Longleftrightarrow\; P_1, \ldots, P_n, Q \vdash \perp$.

Exercise 7.18 $\oslash (P \triangleright Q) \vdash \oslash Q$

Exercise 7.19 $\oslash \oslash (P \triangleright Q) \vdash (P \triangleright \oslash \oslash Q)$. (*Note that* $P, \oslash Q \vdash \oslash (P \triangleright Q)$.)

You have obtained three versions of CONTRAPOSITION (Exercises 7.9, 7.10, and 7.11) from the first of which you can easily obtain the principle of *modus tollens*:

$$(P \triangleright Q), \oslash Q \vdash \oslash P.$$

You have shown that a statement must be absurd if it yields its own absurdity (Exercise 7.12). You have shown that the absurdity of everything follows from absurdity (Exercise 7.13). You have confirmed the principle of DOUBLE NEGATION INTRODUCTION (Exercise 7.14). You have obtained two versions of DOUBLE NEGATION ELIMINATION (Exercises 7.15 and 7.16). And more (Exercises 7.17–7.19). Strangely enough, you have done all this without using EFQ. But EFQ is our only information about what the sentence '$\perp$' might say. In the absence of EFQ, '$\perp$' could assert a logical truth rather than an absurdity. Indeed, '$\perp$' could then say *anything*.

If we want '$\oslash P$' to express the *rejection* or *negation* of 'P', then it matters tremendously what '$\perp$' says. (Noting that a logical truth can be inferred from 'P' is not an effective way of expressing disapproval of 'P'.) It may be surprising, then, that so many properties thought to be characteristic of negation follow from Definition 7.1 and facts about '$\triangleright$'.[5]

From now on, feel free to use EFQ.

---

[5] This is just one of the many insights to be found in Martin [6]. I will be borrowing exercises from that book for the rest of this chapter.

Exercise 7.20 $P, \oslash P \vdash Q$.

Exercise 7.21 $\oslash(P \rhd Q) \vdash \oslash \oslash P$.

Exercise 7.22 $(\oslash Q \rhd \oslash P) \vdash \oslash \oslash (P \rhd Q)$.

Exercise 7.23 $(P \rhd \oslash \oslash Q) \vdash \oslash \oslash (P \rhd Q)$. *(Recall Exercise 7.11.)*

Exercise 7.24 $\oslash(P \mathbin{\&} \oslash Q) \vdash \oslash \oslash (P \rhd Q)$. *(Note Exercise 7.23.)*

7.5 Absurd Absurdities

Suppose it would be absurd for ϕ to be absurd: that is, suppose $\oslash \oslash \phi$. Can we safely infer ϕ itself? Exercises 7.15 and 7.16 showed that this inference is sometimes warranted. Is it *always* warranted? Well, under what circumstances would the absurdity of ϕ trivialize the Peregrinese practice of inference by allowing the Peregrins to infer everything? (This is what it would mean for ϕ's absurdity to be absurd.) One such circumstance is that ϕ is a Peregrinese THEOREM, something they accept without argument. If

$$\emptyset \vdash \phi$$

then, by Exercise 7.14 and Cut,

$$\emptyset \vdash \oslash \oslash \phi.$$

It would be absurd for a theorem to be absurd. Is that the *only* situation in which it is absurd for something to be absurd? Could we have a reason to assert the absurdity of a sentence ϕ's absurdity without having a reason to assert ϕ?

It seems hard to imagine such a situation. It may be instructive to consider a case that turns out *not* to be what we are looking for. Suppose the Peregrins have a kind of *generic* statement

Thingamajig is a whatchamacallit

or, more briefly,

■.

The generic statement is like a black box. Since its contents are inaccessible to the Peregrins, whatever they can prove about the black box, they can prove about anything. (That is the sense in which it is generic.) In particular, if they show that the black box is absurd, then they can show that anything is absurd. But *that* would be absurd. Isn't that a good reason for the Peregrins to assert the absurdity of the absurdity of the black box even though they have no reason to assert the black box itself?

We need to look into this more carefully. To start, we need to say exactly what it means for the black box to be generic. Let $\theta(\blacksquare)$ be a sentence in which '$\blacksquare$' occurs.

Let $\theta(\psi)$ be the result of replacing every occurrence of '■' in $\theta(■)$ with an occurrence of the sentence ψ. Then the black box obeys the following principle.

Proposition 7.1 (GEN)

$$\emptyset \vdash \theta(■) \quad \Rightarrow \quad \emptyset \vdash \theta(\psi).$$

If the Peregrins can prove that the mysterious black box has a property, then they can prove that every sentence has that property. We can now show that the Peregrins can prove an absurdity if they can prove the black box.

Proposition 7.2 $\vdash ■ \Rightarrow \vdash \bot$.

Proof The proof only takes two lines.

> 1. $\vdash ■$ Prem
> 2. $\vdash \bot$ 1 Gen

The Peregrins can also prove an absurdity if they can prove the absurdity of the black box.

Exercise 7.25 $\vdash \oslash ■ \Rightarrow \vdash \bot$.

It might be tempting to treat the preceding result as a proof that it would be absurd for the black box to be absurd:

$$\vdash \oslash \oslash ■$$

or, equivalently,

$$\oslash ■ \vdash \bot .$$

The next result shows that this is correct only if the Peregrinese system of inference allows them to prove absurdities.

Exercise 7.26 $\vdash \oslash \oslash ■ \Rightarrow \vdash \bot$.

In general terms, the situation is this. If a certain sentence ϕ is a Peregrinese theorem, then so is a certain sentence ψ

$$\emptyset \vdash \phi \quad \Rightarrow \quad \emptyset \vdash \psi.$$

(Exercise 7.25 says: $\vdash \oslash ■ \Rightarrow \vdash \bot$.) Yet, if the Peregrinese system is consistent, the Peregrins cannot verify the inference from ϕ to ψ; that is, they cannot verify

$$\phi \vdash \psi.$$

(They cannot verify $\oslash ■ \vdash \bot$.) As we saw earlier in connection with Int, the inference from ϕ to ψ is not the same as the inference from the theoremhood of ϕ to the theoremhood of ψ. When Peregrins ask you to assume a premise for the purpose of deriving a conclusion, they are not asking you to assume that the premise is a theorem for the sake of showing that the conclusion is a theorem. Assuming a premise is not the same as assuming that the premise is provable. That is, it is not intrinsic to the very concept of inference that these assumptions are interchangeable.

Our hope was that a sentence whose status is, in principle, indeterminate would be an example of a sentence whose absurdity is demonstrably absurd even though the sentence itself is not provable. This did not work out. No need to despair, however. The following theorem introduces a useful technique for deriving the absurdity of a sentence's absurdity without deriving the sentence itself. The general idea is that if ψ is derivable from both ϕ and $\oslash\phi$, then it would be absurd for ψ to be absurd because the absurdity of ψ would (by contraposition) yield both $\oslash\phi$ and $\oslash\oslash\phi$.

Theorem 7.10 $\varPhi \cup \{\phi\} \vdash \psi, \ \varPsi \cup \{\oslash\phi\} \vdash \psi \ \Rightarrow \ \varPhi \cup \varPsi \vdash \oslash\oslash\psi.$

Proof See the formal derivation below.

1.	$\varPhi \cup \{\phi\} \vdash \psi$	Prem
2.	$\varPsi \cup \{\oslash\phi\} \vdash \psi$	Prem
3.	$\varPhi \cup \{\oslash\psi\} \vdash \oslash\phi$	1 Ex 7.10
4.	$\varPsi \cup \{\oslash\psi\} \vdash \oslash\oslash\phi$	2 Ex 7.10
5.	$\oslash\phi, \oslash\oslash\phi \vdash \bot$	Ex 7.20
6.	$\varPhi \cup \{\oslash\psi, \oslash\oslash\phi\} \vdash \bot$	3,5 Cut
7.	$\varPhi \cup \varPsi \cup \{\oslash\psi\} \vdash \bot$	4,6 Cut
8.	$\varPhi \cup \varPsi \vdash \oslash\oslash\psi$	7 Ex 7.17

We will now use Theorem 7.10 to prove the absurdity of the absurdity of the following instance of PEIRCE'S LAW.[6]

$$(((P \rhd Q) \rhd P) \rhd P)$$

This sentence is not a theorem of our formal system nor would anything we know about the Peregrins lead us to think that it ought to be.[7] Nonetheless, we now prove its double negation.

Theorem 7.11 $\vdash \oslash\oslash(((P \rhd Q) \rhd P) \rhd P).$

Proof See the formal derivation below.

---

[6] CHARLES SANDERS PEIRCE (1839–1914) was an American logician, mathematician, philosopher, and scientist.

[7] If you can infer P from the inferability of Q from P, can you safely infer P? That move may be justified; but its justifiability is not intrinsic to the very concept of inference. By the way, it is not *obvious* that '$(((P \rhd Q) \rhd P) \rhd P)$' is unprovable in our system. There are proofs of its unprovability, but I will not supply any here. If you want to pursue this further, you might consult the WIKIPEDIA article on Heyting algebras (http://en.wikipedia.org/wiki/Heyting_algebra).

1. $\qquad P \vdash (((P \rhd Q) \rhd P) \rhd P)$ Int
2. $\qquad P, \oslash P \vdash Q$ Ex 7.20
3. $\qquad \oslash P \vdash (P \rhd Q)$ 2 Thm 7.5
4. $(P \rhd Q), ((P \rhd Q) \rhd P) \vdash P$ MP
5. $\qquad \oslash P, ((P \rhd Q) \rhd P) \vdash P$ 3,4 Cut
6. $\qquad \oslash P \vdash (((P \rhd Q) \rhd P) \rhd P)$ 5 Thm 7.5
7. $\qquad \vdash \oslash \oslash (((P \rhd Q) \rhd P) \rhd P)$ 1,6 Thm 7.10

We can learn an important lesson from Theorem 7.11. If each instance of the double negation elimination priniciple

$$\oslash \oslash \phi \vdash \phi$$

were provable, then Theorem 7.11 would yield the instance of Peirce's law that we know to be unprovable. Exercises 7.15 and 7.16 showed that it is sometimes permissible to erase pairs of $\oslash$'s. We now see that such inferences are not always accepted in the land of Peregrin.

Exercise 7.27 *Show that if every instance of Peirce's law is a Peregrinese theorem, then:* $\oslash \oslash P \vdash P$. *(You need to find a formula of the form* $(((\phi \rhd \psi) \rhd \phi) \rhd \phi)$ *that yields the desired result.)*

Exercise 7.28 $\vdash \oslash \oslash (\oslash \oslash P \rhd P)$.

Exercise 7.29 $((P \rhd Q) \rhd P) \vdash \oslash \oslash P$.

Some instances of Peirce's law are Peregrinese theorems. For example, if ϕ is any Peregrinese theorem, then Int guarantees that

$$(((\phi \rhd \psi) \rhd \phi) \rhd \phi)$$

is a Pergrinese theorem no matter what ψ is. Here is another example that you can verify without using EFQ.

Exercise 7.30 $\vdash (((\oslash \oslash P \rhd \bot) \rhd \oslash \oslash P) \rhd \oslash \oslash P)$. *(You might want to use Exercises 7.12 and 7.14 replacing 'P' with a sentence more useful to us now.)*

7.6 Disjunctions

Suppose the Peregrins introduce an expression '$\triangledown$' that behaves as follows.[8]

---

[8] I am not going to try to present '$\triangledown$' as a natural outgrowth of Peregrinese reflections on a logic-free conception of inference. Perhaps you can figure out a way to do so. You might start by looking into something called "multiple-conclusion logic."

Axiom 7.11 (OR)

$$\Phi \cup \{\phi \bigtriangledown \psi\} \vdash \chi \quad \Longleftrightarrow \quad \Phi \cup \{\phi\} \vdash \chi \text{ and } \Phi \cup \{\psi\} \vdash \chi.$$

Students of logic may notice that the right-left direction of Or is an inference form known as SEPARATION OF CASES.

Exercise 7.31 $(P \bigtriangledown P) \vdash P$.

Exercise 7.32 $P \vdash (P \bigtriangledown Q)$.

Theorem 7.12 $Q \vdash (P \bigtriangledown Q)$.

Exercise 7.33 $\vdash \oslash \oslash (P \bigtriangledown \oslash P)$.

Exercise 7.34 *The* LAW OF EXCLUDED MIDDLE (LEM) *is the principle*

$$\emptyset \vdash (\phi \bigtriangledown \oslash \phi).$$

Show that some instances of LEM *are not provable. (You can show that something is unprovable by showing that it would allow you to prove something you already know to be unprovable. I can report that the addition of Axiom 7.11 does not allow us to prove any previously unprovable instances of Peirce's law.)*

Exercise 7.35 $(P \bigtriangledown Q) \vdash (\oslash P \rhd Q)$.

Exercise 7.36 $(\oslash P \rhd Q) \vdash \oslash \oslash (P \bigtriangledown Q)$.

Exercise 7.37 *Show that some instances of the following scheme are unprovable:*

$$(\oslash \phi \rhd \psi) \vdash (\phi \bigtriangledown \psi).$$

Exercise 7.38 $(\oslash P \,\&\, \oslash Q) \vdash \oslash (P \bigtriangledown Q)$.

Exercise 7.39 $\oslash (P \bigtriangledown Q) \vdash (\oslash P \,\&\, \oslash Q)$.

One of the following two sequents is provable; the other is not.

$$\oslash (\oslash P \bigtriangledown \oslash Q) \vdash (P \,\&\, Q).$$
$$(P \,\&\, Q) \vdash \oslash (\oslash P \bigtriangledown \oslash Q).$$

Exercise 7.40 *Prove the provable one.*

Exercise 7.41 *Show that the unprovable one is unprovable. (If your chosen sequent were provable, then the result of replacing each occurrence of 'Q' with an occurrence of 'P' would also be provable. Show that this new sequent would let us prove something unprovable.)*

One of the following two sequents is provable; the other is not.

$$((P \,\&\, \oslash Q) \rhd R) \vdash (P \rhd (Q \,\triangledown\, R)).$$
$$(P \rhd (Q \,\triangledown\, R)) \vdash ((P \,\&\, \oslash Q) \rhd R).$$

Exercise 7.42 *Prove the provable one.*

Exercise 7.43 *Show that the unprovable one is unprovable. (As in Exercise 7.41, feel free to manipulate your chosen sequent by replacing occurrences of 'P', 'Q', and 'R' with occurrences of other formulas. Then show that your new sequent would let you prove something unprovable.)*

I reported (without proof) that Peirce's law has instances not provable in our deductive system. This allowed us to show that there are unprovable instances of double negation elimination and LEM. These facts may also have helped you do Exercises 7.37, 7.41, and 7.43. This approach does not always work: there are unprovable sentences that do not yield any unprovable instances of Peirce's law. An example is

$$(\oslash P \,\triangledown\, \oslash \oslash P)$$

which is an instance of the PRINCIPLE OF TESTABILITY

$$(\oslash \phi \,\triangledown\, \oslash \oslash \phi).$$

This observation may help you with some of the following exercises.

Exercise 7.44 *Show that some instances of the following scheme are unprovable[9]:*

$$\emptyset \vdash ((\phi \rhd \psi) \,\triangledown\, (\psi \rhd \phi)).$$

One of the following two sequents is provable; the other is not.

$$(\oslash P \,\triangledown\, \oslash Q) \vdash \oslash(P \,\&\, Q).$$
$$\oslash(P \,\&\, Q) \vdash (\oslash P \,\triangledown\, \oslash Q).$$

Exercise 7.45 *Prove the provable one.*

Exercise 7.46 *Show that the unprovable one is unprovable.*

---

[9] Is it intrinsic to the concept of inference that, given any two sentences, one will be inferable from the other?

7.7 Assimilators

To the east of the Peregrins lies the land of Boole. The Boolean language is similar to Peregrinese in many ways. For example, the Booleans have a symbol '$\bot$' that obeys EFQ. They also have a symbol '$\neg$' with some of the logical properties of the Peregrinese '$\oslash$'.

Proposition 7.3

$$\Phi \vdash \neg\phi \iff \Phi \cup \{\phi\} \vdash \bot .$$

Booleans will infer $\neg\phi$ from Φ if and only if they will infer *anything* from $\Phi \cup \{\phi\}$. This is the Boolean version of Exercise 7.17. The Booleans are an uninhibited lot and, unlike their cautious neighbors to the west, make free use of double negation elimination.

Proposition 7.4

$$\{\neg\neg\phi\} \vdash \phi.$$

The most easterly Peregrins had frequent contact with their Boolean neighbors and adopted many of their customs. In particular, they found it useful to incorporate some Boolean elements into their dialect: the expression '$\neg$', for example, with all its logical properties. They soon discovered, to their horror, that they were losing their cultural identity.

Proposition 7.5 $\oslash\oslash P \vdash P.$

Proof Here is a formal derivation.

1.	$\neg P \vdash \neg P$	Taut
2.	$\neg P,\ P \vdash \bot$	1 Prop 7.3
3.	$\neg P \vdash \oslash P$	2 Thm 7.5
4.	$\oslash\oslash P,\ \oslash P \vdash \bot$	Ex 7.20
5.	$\oslash\oslash P,\ \neg P \vdash \bot$	3,4 Cut
6.	$\oslash\oslash P \vdash \neg\neg P$	5 Prop 7.3
7.	$\neg\neg P \vdash P$	Prop 7.4
8.	$\oslash\oslash P \vdash P$	6,7 Cut

The Peregrins were just trying to be nice. They thought they could ease relations with their neighbors by adopting some Boolean customs while holding tight to their own sacred traditions. Alas, Boolean logic does not tolerate difference. It does not coexist: it assimilates.[10] (Resistance is futile.) Add Boolean negation to Peregrinese logic and the result is a copy of Boolean logic in which '$\oslash$' behaves just like '$\neg$'. (In more conventional terminology: add classical negation to intuitionist logic and the result is classical logic.) '$\triangleright$' begins to act strangely too.

---

[10] For more on assimilators, see Pollard [8–10]. The first two papers are available via http://www. projecteuclid.org. You can find the third at http://www.philosophy.unimelb.edu.au/ajl/

Proposition 7.6 $\vdash (((P \rhd Q) \rhd P) \rhd P)$.

Proof Just apply Theorem 7.11 and Proposition 7.5.

Exercise 7.47 *In the land of Peregrin, there is a monastic community whose members may only utter Π_1^0 sentences and compounds of Π_1^0 sentences formed with '&', '$\rhd$', '$\bot$', '$\oslash$', and '$\triangledown$'. The monks infer '$\bot$' from '$| = ||$'. They infer $--\phi$ from $(-\phi \rhd \bot)$. They infer '$f(\mathfrak{a}) = |$' from '$id(f(\mathfrak{a}), |) = ||$'. Without necessarily giving a full-blown proof, indicate why the monks will infer ϕ from $\oslash \oslash \phi$ whenever ϕ is a Π_1^0 sentence.*

To the west of the Peregrins lies the land of Peirce. The Peirceans have a symbol '$\supset$' with some of the logical properties of '$\rhd$'. Here is the Peircean version of Theorem 7.7.

Proposition 7.7

$$\Phi \cup \{\phi\} \vdash \psi \iff \Phi \vdash (\phi \supset \psi).$$

A very un-Peregrinese trait, however, is the Peircean's acceptance of every instance of Peirce's law.

Proposition 7.8

$$\emptyset \vdash (((\phi \supset \psi) \supset \phi) \supset \phi).$$

Perhaps you can guess what happens when the neighborly Peregrins absorb '$\supset$' into their dialect.

Exercise 7.48 $(P \supset Q), P \vdash Q$.

Proposition 7.9 $(P \supset Q) \vdash (P \rhd Q)$.

Proof Just apply Exercise 7.48 and Theorem 7.5.

Exercise 7.49 $((P \rhd Q) \rhd P) \vdash ((P \supset Q) \supset P)$.

Proposition 7.10 $((P \supset Q) \supset P) \vdash P$.

Proof Just apply Propositions 7.7 and 7.8.

Exercise 7.50 $\vdash (((P \rhd Q) \rhd P) \rhd P)$.

Peircean logic does not tolerate difference. It does not coexist: it assimilates. Add the Peircean conditional to Peregrinese logic and the result is a copy of Peircean logic in which '$\rhd$' behaves just like '$\supset$'. (In more conventional terminology: add the classical conditional to intuitionist logic and the result is classical logic.) '$\oslash$' begins to act strangely too.

Proposition 7.11 $\oslash \oslash P \vdash P$.

Proof Apply Exercises 7.27 and 7.50.

To the south of the Peregrins lies the land of Trivalence. The Trivalents recognize one form of truth (T) but two forms of untruth (U_1 and U_2). They attribute T to a sentence ϕ by asserting ϕ itself. They attribute U_1 to ϕ by asserting $\twoheadrightarrow \phi$. They attribute U_2 to ϕ by asserting $\twoheadrightarrow\twoheadrightarrow \phi$. Since neither form of untruth is compatible with truth, the Trivalents enjoy two versions of EFQ.

Proposition 7.12

$$\{\phi, \; \twoheadrightarrow \phi\} \vdash \psi.$$
$$\{\phi, \; \twoheadrightarrow\twoheadrightarrow \phi\} \vdash \psi.$$

Since a declarative Trivalent sentence has to be in one of the three aforementioned states (T, U_1, U_2), exactly one of the sentences

$$\phi, \; \twoheadrightarrow \phi, \; \twoheadrightarrow\twoheadrightarrow \phi$$

will be true. So anything that follows from each of these three sentences will be true. More generally, if you are able to infer ψ whenever you add one of these sentences to Φ, then you can infer ψ from Φ alone.

Proposition 7.13

$$\Phi \cup \{\phi\} \vdash \psi, \quad \Phi \cup \{\twoheadrightarrow \phi\} \vdash \psi, \quad \Phi \cup \{\twoheadrightarrow\twoheadrightarrow \phi\} \vdash \psi \;\; \Rightarrow \;\; \Phi \vdash \psi.$$

Any guesses about what happens when the most southerly Peregrins extend the hand of good fellowship and start using '$\twoheadrightarrow$'?

Exercise 7.51 $\oslash \oslash P \vdash \oslash \twoheadrightarrow P$.

Exercise 7.52 $\oslash \oslash P \vdash \oslash \twoheadrightarrow\twoheadrightarrow P$.

Exercise 7.53 $\oslash \oslash P \vdash P$.

Need I say it? '$\twoheadrightarrow$' is an assimilator.

Exercise 7.54 *The Peregrins have neighbors to the north who recognize one form of truth and three forms of untruth. You complete the story.*

Classical (two-valued) logic does not play nice. Neither does n-valued logic for any finite n.

7.8 The Glivenko/Gödel Theorems

We now bid adieu to the Peregrins and consider a formal language with the following vocabulary.

1. Three CONNECTIVES: '&', '$\triangleright$', and '$\triangledown$'.
2. One SENTENTIAL CONSTANT: '$\perp$'.
3. Infinitely many SENTENCE LETTERS: 'P', 'Q', 'P_1', 'Q_1',...
4. Two PARENTHESES: '(', ')'.

We recursively define the SENTENCES of our language as follows.

5. '$\perp$' is a sentence.
6. Every sentence letter is a sentence.
7. If ϕ and ψ are sentences, then so are $\ulcorner(\phi \,\&\, \psi)\urcorner$, $\ulcorner(\phi \triangleright \psi)\urcorner$, and $\ulcorner(\phi \triangledown \psi)\urcorner$.

We introduce '$\oslash$' via Definition 7.1. $\vdash_I$ will be the relation governed by Axioms 7.1–7.11. (The 'I' stands for "intuitionist.") $\vdash_C$ will be the relation governed by Axioms 7.1 − 7.11 *and* LEM. (The 'C' stands for "classical.") A derivation of a sequent $\Phi \vdash_I \psi$ or $\Phi \vdash_C \psi$ will have the same form as those in the preceding sections: a series of numbered lines each featuring a sequent and a justification. In a classical derivation, we accept LEM as a justification for any sequent of the form $\emptyset \vdash_C (\phi \triangledown \oslash\phi)$. Feel free to use results we have already verified. Since every step in an intuitionist derivation is a correct step in a classical derivation, we obtain the following theorem.

Theorem 7.13 *If there is an intuitionist derivation of $\Phi \vdash_I \psi$, then there is a classical derivation of $\Phi \vdash_C \psi$.*

The next three exercises will help us verify two theorems first proved by VALERII GLIVENKO (1897–1940).[11]

Exercise 7.55 $(P \triangleright \oslash Q), \oslash\oslash P \vdash_I \oslash Q$.

Exercise 7.56 $((P \triangledown \oslash P) \triangleright \oslash Q) \vdash_I \oslash Q$.

Exercise 7.57 $\Phi \cup \{(\phi \triangledown \oslash\phi)\} \vdash_I \oslash\psi \;\Rightarrow\; \Phi \vdash_I \oslash\psi$.

Here is a close cousin of Exercise 7.57.

Exercise 7.58 $\Phi \cup \{(\phi \triangledown \oslash\phi)\} \vdash_I \psi \;\Rightarrow\; \Phi \vdash_I \oslash\oslash\psi$.

The only distinctively classical features of one of our classical derivations will be lines invoking LEM. Exercise 7.57 implies that every such line is eliminable if our conclusion is a negation. That is, we have a technique for transforming a classical derivation of a negation into an intuitionist derivation. The following result will help us confirm this.

---

[11] See Glivenko [2]; English translation in Mancosu [5, pp. 301–305].

Suppose we are presented with a classical derivation of the sequent $\varXi \vdash_C \psi$. Each application of LEM in the derivation will introduce a sequent $\emptyset \vdash_C (\phi \triangledown \oslash \phi)$. Let the members of $\varUpsilon$ be the sentences $(\phi \triangledown \oslash \phi)$ on the right-hand sides of those sequents. We claim there is an intuitionist derivation of the sequent $\varXi \cup \varUpsilon' \vdash_I \psi$ for some subset $\varUpsilon'$ of $\varUpsilon$. Our proof is by induction on the length of the classical derivation. That is, having confirmed that derivations of length 1 satisfy our claim, we will show that derivations of length $n+1$ will satisfy our claim as long as derivations of length n do. Suppose the derivation consists of just one line. If that line features an instance of LEM, $\emptyset \vdash_C (\phi \triangledown \oslash \phi)$, then we just note that, by Taut, there is an intuitionist derivation of $\{(\phi \triangledown \oslash \phi)\} \vdash_I (\phi \triangledown \oslash \phi)$. The only other possible occupants of the one line would be instances of the intuitionist Axioms 7.1, 7.6, 7.7, 7.8, 7.9, or 7.10. (All the other axioms involve a derivation from a previous line.) Now suppose $\varXi \vdash_C \psi$ sits at the end of a derivation with more than one line. The axioms we just discussed present no new problems. So we need only consider Axioms 7.2, 7.3, 7.4, 7.5, and 7.11. Suppose the last line of the classical derivation features an application of Axiom 7.2 (Thin). Then $\varXi$ is of the form $\varPhi \cup \{\phi\}$ and the sequent $\varPhi \vdash_C \psi$ appears on an earlier line. By our inductive hypothesis, there is an intuitionist derivation of $\varPhi \cup \varUpsilon' \vdash_I \psi$ for some subset $\varUpsilon'$ of $\varUpsilon$. So, by Thin, there is an intuitionist derivation of $\varPhi \cup \{\phi\} \cup \varUpsilon' \vdash_I \psi$. That is, there is an intuitionist derivation of $\varXi \cup \varUpsilon' \vdash_I \psi$, as desired. Now suppose the last line of the classical derivation features an application of Axiom 7.3 (Cut). Then $\varXi$ is of the form $\varPhi \cup \varPsi$ and sequents $\varPsi \vdash_C \chi$ and $\varPhi \cup \{\chi\} \vdash_C \psi$ appear on earlier lines. By our inductive hypothesis, there are intuitionist derivations of $\varPsi \cup \varUpsilon' \vdash_I \chi$ and $\varPhi \cup \{\chi\} \cup \varUpsilon'' \vdash_I \psi$ for some subsets $\varUpsilon'$ and $\varUpsilon''$ of $\varUpsilon$. So, by Cut, there is an intuitionist derivation of $\varPhi \cup \varPsi \cup \varUpsilon' \cup \varUpsilon'' \vdash_I \psi$ where $\varUpsilon' \cup \varUpsilon''$ is a subset of $\varUpsilon$, as desired. Suppose the last line of the classical derivation features an application of the left-right direction of Axiom 7.4 (And). Then $\varXi$ is of the form $\{(\phi_1 \& \phi_2)\}$ and the sequent $\{\phi_1, \phi_2\} \vdash_C \psi$ appears on an earlier line. By our inductive hypothesis, there is an intuitionist derivation of $\{\phi_1, \phi_2\} \cup \varUpsilon' \vdash_I \psi$ for some subset $\varUpsilon'$ of $\varUpsilon$. So, by Theorem 7.3, there is an intuitionist derivation of $\{(\phi_1 \& \phi_2)\} \cup \varUpsilon' \vdash_I \psi$. Exercise 7.2 lets us handle the left-right direction of Axiom 7.4 in a similar way. I leave it to you to show how to handle applications of Axioms 7.5 and 7.11.

Theorem 7.14 *If there is a classical derivation of $\varPhi \vdash_C \oslash \psi$, then there is an intuitionist derivation of $\varPhi \vdash_I \oslash \psi$.*

Proof Suppose we are presented with a classical derivation of $\varPhi \vdash_C \oslash \psi$. Then, as we just showed, there is an intuitionist derivation of $\varPhi \cup \varUpsilon \vdash_I \oslash \psi$ where $\varUpsilon$ is a finite set of sentences of the form $(\phi \triangledown \oslash \phi)$. Finitely many applications of Exercise 7.57 yield an intuitionist derivation of $\varPhi \vdash_I \oslash \psi$. (If you have any doubts, you can use induction on the size of $\varUpsilon$ to confirm this last claim.)

Theorem 7.15 *If there is a classical derivation of $\varPhi \vdash_C \psi$, then there is an intuitionist derivation of $\varPhi \vdash_I \oslash \oslash \psi$.*

Proof If there is a classical derivation of $\varPhi \vdash_C \psi$, then, by Exercise 7.14, there is a classical derivation of $\varPhi \vdash_C \oslash \oslash \psi$. Now apply Theorem 7.14.

We will now see how Kurt Gödel improved on Glivenko's results.[12] First some definitions.

Definition 7.2 ϕ is an INTUITIONIST THEOREM if and only if there is an intuitionist derivation of $\emptyset \vdash_I \phi$.

Definition 7.3 ϕ is a CLASSICAL THEOREM if and only if there is a classical derivation of $\emptyset \vdash_C \phi$.

Definition 7.4 ϕ and ψ are INTUITIONISTICALLY EQUIVALENT if and only if there are intuitionist derivations of $\{\phi\} \vdash_I \psi$ and $\{\psi\} \vdash_I \phi$.

Definition 7.5 ϕ and ψ are CLASSICALLY EQUIVALENT if and only if there are classical derivations of $\{\phi\} \vdash_C \psi$ and $\{\psi\} \vdash_C \phi$.

Definition 7.6 A $\{\oslash, \&\}$-SENTENCE is a sentence in which no connectives other than '$\oslash$' and '$\&$' occur.

I note two facts without proof. First, neither '$\bot$' nor any sentence letter is a classical theorem. Second, every sentence of our formal language is classically equivalent to a $\{\oslash, \&\}$-sentence.

Definition 7.7 The LENGTH of a sentence is the number of occurrences of connectives within it.

For example, '$\bot$' and 'P' have length 0, while '$\oslash \oslash (P \triangledown (P \triangleright Q))$' has length 4 (with two occurrences of '$\oslash$' and one occurrence each of '$\triangledown$' and '$\triangleright$').

Theorem 7.16 *A $\{\oslash, \&\}$-sentence is a classical theorem only if it is an intuitionist theorem.*

Proof Our proof will be by induction on the length of our sentences. That is, having confirmed that sentences of length 0 satisfy the theorem, we will show that sentences of length $n + 1$ will satisfy the theorem as long as sentences of length n do. Since no sentence of length 0 is a classical theorem, every sentence of length 0 trivially satisfies the theorem. A $\{\oslash, \&\}$-sentence with non-zero length will be either a negation or a conjunction. Theorem 7.14 takes care of the negations. In the case of a conjunction $(\phi \& \psi)$, our inductive hypothesis is that both ϕ and ψ satisfy the theorem. Suppose $(\phi \& \psi)$ is a classical theorem. Then ϕ and ψ are classical theorems and, hence, ϕ and ψ are intuitionist theorems. It follows, as desired, that $(\phi \& \psi)$ is an intuitionist theorem.

We now define an INTERPRETATION FUNCTION i as follows. $i(\phi) = \phi$ whenever ϕ is '$\bot$' or a sentence letter. Furthermore:

---

[12] See Gödel [3]; English translation in Gödel [4, pp. 287–295].

$$i(\phi \,\&\, \psi) = (i(\phi) \,\&\, i(\psi))$$
$$i(\phi \rhd \psi) = \oslash(i(\phi) \,\&\, \oslash i(\psi))$$
$$i(\phi \bigtriangledown \psi) = \oslash(\oslash i(\phi) \,\&\, \oslash i(\psi)).$$

Note that

$$i(\oslash\phi) = i(\phi \rhd \bot) = \oslash(i(\phi) \,\&\, \oslash \,\bot).$$

You can use Theorems 7.1 and 7.9 and Exercise 7.10 to derive $\oslash(i(\phi) \,\&\, \oslash \,\bot)$ from $\oslash i(\phi)$ and vice versa. So $i(\oslash\phi)$ and $\oslash i(\phi)$ are intuitionistically and classically equivalent. It follows that you can interchange these sentences without affecting derivability. That is, if you replace occurrences of one sentence with occurrences of the other in an intuitionistically or classically derivable sequent, the result will be an intuitionistically or classically derivable sequent.[13] Since $i(\oslash\phi)$ and $\oslash i(\phi)$ are interchangeable, it is harmless to treat them as identical:

$$i(\oslash\phi) = \oslash i(\phi).$$

The function i transforms every sentence into a classically equivalent $\{\oslash, \&\}$-sentence. For example,

$$i(\oslash \oslash P \rhd P) = \oslash(\oslash \oslash P \,\&\, \oslash P).$$

This lets us show that i assigns an intuitionist theorem to every classical theorem.

Theorem 7.17 *If ψ is a classical theorem, then $i(\psi)$ is an intuitionist theorem.*

Proof If ψ is a classical theorem, then so is $i(\psi)$ (since ψ and $i(\psi)$ are classically equivalent). But then, by Theorem 7.16, $i(\psi)$ is an intuitionist theorem.

No classical derivation of a theorem is pointless from an intuitionist perspective. Every such derivation yields an intuitionist theorem. If the classical theorem is a $\{\oslash, \&\}$-sentence, then, by Theorem 7.16, that sentence is itself an intuitionist theorem. In every other case, we can apply Theorem 7.15 or 7.17.

7.9 First-Order Logic With a Decidable Relation

We will now confirm that a similar result applies to a much more expressive formal language. Our new language features the following vocabulary.

1. Three CONNECTIVES: '&', '$\rhd$', and '$\bigtriangledown$'.
2. One QUANTIFIER: '$\forall$'.
3. One RELATION symbol: 'R'.
4. One SENTENTIAL CONSTANT: '$\bot$'.

---

[13] For a proof, see Martin [6, p. 168].

5. Infinitely many VARIABLES: 'w', 'x', 'y', 'z', 'w_1', 'x_1', 'y_1', 'z_1',...
6. TWO PARENTHESES: '(', ')'.

We recursively define the FORMULAS of our language as follows.

7. '$\bot$' is a formula.
8. If α and β are variables, then $\ulcorner R\alpha\beta \urcorner$ is a formula.
9. If ϕ and ψ are formulas, then so are $\ulcorner (\phi \mathbin{\&} \psi) \urcorner$, $\ulcorner (\phi \rhd \psi) \urcorner$, and $\ulcorner (\phi \bigtriangledown \psi) \urcorner$.
10. If α is a variable and ϕ is a formula, then $\ulcorner \forall\alpha\,\phi \urcorner$ is a formula.

We introduce '$\oslash$' via Definition 7.1. We define FREE and BOUND occurrences of variables as in Chap. 2.

Axiom 7.12 If $\phi(\beta)$ is the result of replacing every free occurrence of the variable α in $\phi(\alpha)$ with a free occurrence of the variable β, then

$$\{\forall\alpha\,\phi(\alpha)\} \vdash \phi(\beta).$$

If $\alpha = \beta$, this takes the form

$$\{\forall\alpha\,\phi\} \vdash \phi.$$

Axiom 7.13 If the variable α does not occur free in any member of $\varPhi$, then

$$\varPhi \vdash \phi \quad \Rightarrow \quad \varPhi \vdash \forall\alpha\,\phi.$$

Axiom 7.14

$$\emptyset \vdash \forall x \forall y (Rxy \bigtriangledown \oslash Rxy).$$

Axiom 7.14 says that the relation R satisfies a version of **LEM**. Such a relation is said to be DECIDABLE. Now let $\vdash_I$ be the relation governed by Axioms 7.1–7.14. Let $\vdash_C$ be the relation governed by Axioms 7.1–7.13 *and* **LEM**. Since Axiom 7.14 is a classical theorem, we once again have:

$$\varPhi \vdash_I \psi \quad \Rightarrow \quad \varPhi \vdash_C \psi.$$

We extend our interpretation function i to our new language by stipulating:

$$i(R\alpha\beta) = R\alpha\beta$$
$$i(\forall\alpha\,\phi) = \forall\alpha\,i(\phi).$$

When we assess the length of a formula we will now count occurrences of the quantifier '$\forall$' as well as occurrences of connectives. For example, '$\forall x \oslash \forall y \oslash Rxy$' has length 4.

Exercise 7.59 $\emptyset \vdash_I \forall x \forall y (\oslash \oslash Rxy \rhd Rxy)$.

Exercise 7.60 $\oslash \oslash (\phi \mathbin{\&} \psi) \vdash_I (\oslash \oslash \phi \mathbin{\&} \oslash \oslash \psi)$.

Exercise 7.61 $\{\phi\} \vdash_I \psi \;\; \Rightarrow \;\; \{\forall \alpha \, \phi\} \vdash_I \forall \alpha \, \psi$.

Exercise 7.62 $\{\oslash \oslash \forall \alpha \, \phi\} \vdash_I \forall \alpha \, \oslash \oslash \phi$.

Exercise 7.63 *Use induction on the length of the formula $i(\phi)$ to show:*

$$\{\oslash \oslash i(\phi)\} \vdash_I i(\phi).$$

You might find Exercise 7.15 useful.

Exercise 7.64 $\{(i(\phi) \rhd i(\psi))\} \vdash_I i(\phi \rhd \psi)$.

Exercise 7.65 $\{i(\phi \rhd \psi)\} \vdash_I (i(\phi) \rhd i(\psi))$. *(You might take a look at Exercises 7.19 and 7.24.)*

Exercise 7.66 $\{i(\phi)\} \vdash_I i(\phi \triangledown \psi)$ *and* $\{i(\psi)\} \vdash_I i(\phi \triangledown \psi)$.

Exercise 7.67

$$\Phi \cup \{i(\phi)\} \vdash_I i(\chi) \;\; and \;\; \Phi \cup \{i(\psi)\} \vdash_I i(\chi) \;\; \Rightarrow \;\; \Phi \cup \{i(\phi \triangledown \psi)\} \vdash_I i(\chi).$$

Exercise 7.68 *Show that if*

$$\phi_1, \ldots, \phi_n \vdash_C \psi$$

is an instance of a classical axiom, then there is an intuitionist derivation of

$$i(\phi_1), \ldots, i(\phi_n) \vdash_I i(\psi).$$

You will need to check all the classical axioms that allow you to write down a sequent without any help from earlier sequents. They are: 7.1, 7.6, 7.7, 7.8, 7.9, 7.10, 7.12, and **LEM**. *You will probably find Exercises 7.64 and 7.65 useful.*

Exercise 7.69 *Suppose we have a classical derivation of $\{\phi_1, \ldots, \phi_n\} \vdash_C \psi$. Use induction on the length of that derivation to show that there is an intuitionist derivation of $\{i(\phi_1), \ldots, i(\phi_n)\} \vdash_I i(\psi)$.*

If we add more relation symbols to our language, then, as long as the corresponding versions of Axiom 7.14 are intuitionist theorems, the preceding results will continue to hold. We might, for example, add symbols meant to express the relations "x is the successor of y," "x is the sum of y and z," and "x is the product of y and z." All three relations are primitive recursive and intuitionists agree that all three are decidable. Let HA be the result of retaining all the axioms of PA but switching the underlying logic from classical to intuitionist.[14] Then the result of applying an interpretation function i with the above properties to an axiom of PA is always a theorem of HA. So, as Gödel was quick to observe, PA is consistent if HA is. For suppose PA is inconsistent. Then, for some PA axioms $\phi_1, \ldots, \phi_n$ and any PA formula ψ,

---

[14] The 'H' in HA commemorates the mathematician Arend Heyting (1898–1980).

$$\{\phi_1, \ldots, \phi_n\} \vdash_C (\psi \& \oslash \psi).$$

Applying a version of Exercise 7.69, we get:

$$\{i(\phi_1), \ldots, i(\phi_n)\} \vdash_I i(\psi \& \oslash \psi).$$

But

$$i(\psi \& \oslash \psi) = (i(\psi) \& \oslash i(\psi)).$$

So HA is inconsistent if PA is.[15] If you accept the consistency of intuitionist arithmetic, you are in no position to question the consistency of PA.

7.10 Curry's Paradox

We conclude this chapter, and the book, with another example of how perilous set theory can be. Suppose a Peregrinese mathematician proposes the following.

Proposition 7.14 (SET)

$$\beta \in \{\alpha : \theta(\alpha)\} \dashv \vdash \theta(\beta).$$

The idea is that β will belong to the set of those things that have property θ if and only if β has property θ. β will belong to the set of prime numbers if and only if β is a prime number. Our experience with lists back in Chap. 3 should make us wary here: recall the non-existent list of all lists that do not list themselves. Let us consider the Peregrinese set of all sets that are not members of themselves: $\{x : x \notin x\}$ or, in unabbreviated notation, $\{x : (x \in x \rhd \bot)\}$. For brevity's sake, let us call this set 'c'. Note that c is a member of itself if and only if $(c \in c \rhd \bot)$.

Proposition 7.15 $\vdash \bot$.

Proof Here is a formal derivation.

1.	$c \in c \vdash (c \in c \rhd \bot)$	Set
2. $c \in c, (c \in c \rhd \bot) \vdash \bot$		MP
3.	$c \in c \vdash \bot$	1,2 Cut
4.	$\vdash (c \in c \rhd \bot)$	3 Ded
5.	$(c \in c \rhd \bot) \vdash c \in c$	Set
6.	$\vdash c \in c$	4,5 Cut
7.	$\vdash \bot$	3,6 Cut

Note that we did not use Int. Skepticism about that rule will not keep you out of trouble here. Even more interesting: we did not use EFQ. Since we used no

---

[15] Again, see Gödel [3]; English translation in Gödel [4, pp. 287–295].

special information about '$\bot$', the proof will still work when we replace '$\bot$' with any sentence. So, given only Cut, Ded, and MP as our logical apparatus, the rule Set allows us to show that every sentence is a theorem.[16]

7.11 Solutions of Odd-Numbered Exercises

7.1

$$
\begin{array}{lll}
1. & \{(\phi_1 \& \phi_2)\} \vdash (\phi_1 \& \phi_2) & \text{Taut} \\
2. & \{\phi_1, \phi_2\} \vdash (\phi_1 \& \phi_2) & \text{1 And}
\end{array}
$$

7.3

$$
\begin{array}{lll}
1. & S_1, P \vdash Q & \text{Prem} \\
2. & S_2, P \vdash R & \text{Prem} \\
3. & Q, R \vdash (Q \& R) & \text{Ex 7.1} \\
4. & S_1, P, R \vdash (Q \& R) & \text{1, 3 Cut} \\
5. & S_1, S_2, P \vdash (Q \& R) & \text{2, 4 Cut}
\end{array}
$$

7.5

$$
\begin{array}{lll}
1. & (P \& (Q \& R)) \vdash P & \text{Thm 7.1} \\
2. & (P \& (Q \& R)) \vdash (Q \& R) & \text{Thm 7.2} \\
3. & (Q \& R) \vdash Q & \text{Thm 7.1} \\
4. & (Q \& R) \vdash R & \text{Thm 7.2} \\
5. & (P \& (Q \& R)) \vdash Q & \text{2, 3 Cut} \\
6. & (P \& (Q \& R)) \vdash R & \text{2, 4 Cut} \\
7. & (P \& (Q \& R)) \vdash (P \& Q) & \text{1, 5 Ex 7.3} \\
8. & (P \& (Q \& R)) \vdash ((P \& Q) \& R) & \text{6, 7 Ex 7.3}
\end{array}
$$

7.7

$$
\begin{array}{lll}
1. & P_1, \ldots, P_n \vdash (Q \rhd R) & \text{Prem} \\
2. & Q, (Q \rhd R) \vdash R & \text{MP} \\
3. & P_1, \ldots, P_n, Q \vdash R & \text{1, 2 Cut}
\end{array}
$$

---

[16] The American logician HASKELL CURRY (1900–1982) noted this in [1] (http://www.jstor.org/stable/2269292).

7.9

1.	$P, (P \rhd Q) \vdash Q$	MP
2.	$Q, (Q \rhd \bot) \vdash \bot$	MP
3. $P, (P \rhd Q), (Q \rhd \bot) \vdash \bot$		1, 2 Cut
4.	$(P \rhd Q), (Q \rhd \bot) \vdash (P \rhd \bot)$	3 Thm 7.6
5.	$(P \rhd Q) \vdash ((Q \rhd \bot) \rhd (P \rhd \bot))$	4 Thm 7.5

7.11

1.	$P, (P \rhd (Q \rhd \bot)) \vdash (Q \rhd \bot)$	MP
2.	$Q, (Q \rhd \bot) \vdash \bot$	MP
3. $P, Q, (P \rhd (Q \rhd \bot)) \vdash \bot$		1, 2 Cut
4.	$Q, (P \rhd (Q \rhd \bot)) \vdash (P \rhd \bot)$	3 Thm 7.6
5.	$(P \rhd (Q \rhd \bot)) \vdash (Q \rhd (P \rhd \bot))$	4 Thm 7.5

7.13

1. $\bot \vdash (P \rhd \bot)$ Int

7.15

1.	$P \vdash \oslash \oslash P$	Ex 7.14
2. $\oslash \oslash \oslash P \vdash \oslash P$		1 Ex 7.10

7.17 Since '$\oslash Q$' is shorthand for '$(Q \rhd \bot)$', this exercise is just an instance of Theorem 7.7.

7.19

1.	$P, (P \rhd Q) \vdash Q$	MP
2.	$P, \oslash Q \vdash \oslash (P \rhd Q)$	1 Ex 7.10
3. $P, \oslash \oslash (P \rhd Q) \vdash \oslash \oslash Q$		2 Ex 7.10
4.	$\oslash \oslash (P \rhd Q) \vdash (P \rhd \oslash \oslash Q)$	3 Thm 7.5

7.21

1.	$P, \oslash P \vdash Q$	Ex 7.20
2.	$\oslash P \vdash (P \rhd Q)$	1 Thm 7.5
3. $\oslash (P \rhd Q) \vdash \oslash \oslash P$		2 Ex 7.10

7.23

> 1. $(P \triangleright \oslash \oslash Q) \vdash (\oslash Q \triangleright \oslash P)$ Ex 7.11
> 2. $(\oslash Q \triangleright \oslash P) \vdash \oslash \oslash (P \triangleright Q)$ Ex 7.22
> 3. $(P \triangleright \oslash \oslash Q) \vdash \oslash \oslash (P \triangleright Q)$ 1, 2 Cut

7.25

> 1. $\vdash \oslash \blacksquare$ Prem
> 2. $\vdash \oslash \oslash \perp$ 1 Gen
> 3. $\oslash \oslash \perp \vdash \perp$ Ex 7.16
> 4. $\vdash \perp$ 2, 3 Cut

7.27 Note that '$((\oslash P \triangleright P) \triangleright P)$' is an instance of Peirce's law because it is shorthand for '$(((P \triangleright \perp) \triangleright P) \triangleright P)$'.

> 1. $\vdash ((\oslash P \triangleright P) \triangleright P)$ Prem
> 2. $\oslash P, \oslash \oslash P \vdash P$ Ex 7.20
> 3. $\oslash \oslash P \vdash (\oslash P \triangleright P)$ 2 Thm 7.5
> 4. $(\oslash P \triangleright P), ((\oslash P \triangleright P) \triangleright P) \vdash P$ MP
> 5. $(\oslash P \triangleright P) \vdash P$ 1, 4 Cut
> 6. $\oslash \oslash P \vdash P$ 3, 5 Cut

7.29

> 1. $\oslash (P \triangleright Q) \vdash \oslash \oslash P$ Ex 7.21
> 2. $(P \triangleright Q), ((P \triangleright Q) \triangleright P) \vdash P$ MP
> 3. $P \vdash \oslash \oslash P$ Ex 7.14
> 4. $(P \triangleright Q), ((P \triangleright Q) \triangleright P) \vdash \oslash \oslash P$ 2, 3 Cut
> 5. $((P \triangleright Q) \triangleright P) \vdash \oslash \oslash P$ 1, 4 Thm 7.10

7.31

> 1. $P \vdash P$ Taut
> 2. $(P \triangledown P) \vdash P$ 1 Or

7.33

> 1. $P \vdash (P \triangledown \oslash P)$ Ex 7.32
> 2. $\oslash P \vdash (P \triangledown \oslash P)$ Thm 7.12
> 3. $\vdash \oslash \oslash (P \triangledown \oslash P)$ 1, 2 Thm 7.10

7.35

1.	$P, \oslash P \vdash Q$	Ex 7.20
2.	$P \vdash (\oslash P \rhd Q)$	1 Thm 7.5
3.	$Q \vdash (\oslash P \rhd Q)$	2 Int
4.	$(P \triangledown Q) \vdash (\oslash P \rhd Q)$	2, 3 Or

7.37

1.	$\vdash (\oslash P \rhd \oslash P)$	Ex 7.6
2.	$(\oslash P \rhd \oslash P) \vdash (P \triangledown \oslash P)$	Prem
3.	$\vdash (P \triangledown \oslash P)$	1, 2 Cut

7.39

1.	$P \vdash (P \triangledown Q)$	Ex 7.32
2.	$Q \vdash (P \triangledown Q)$	Thm 7.12
3.	$\oslash(P \triangledown Q) \vdash \oslash P$	1 Ex 7.10
4.	$\oslash(P \triangledown Q) \vdash \oslash Q$	2 Ex 7.10
5.	$\oslash(P \triangledown Q) \vdash (\oslash P \& \oslash Q)$	3, 4 Ex 7.3

7.41 By Theorem 7.11, we can prove every instance of Peirce's law if we can prove: $\oslash \oslash P \vdash P$. Now note the following.

1.	$(\oslash \oslash P \& \oslash \oslash P) \vdash \oslash(\oslash P \triangledown \oslash P)$	Ex 7.38
2.	$\oslash \oslash P \vdash (\oslash \oslash P \& \oslash \oslash P)$	Ex 7.1
3.	$\oslash \oslash P \vdash \oslash(\oslash P \triangledown \oslash P)$	1, 2 Cut
4.	$\oslash(\oslash P \triangledown \oslash P) \vdash (P \& P)$	Prem
5.	$\oslash \oslash P \vdash (P \& P)$	3, 4 Cut
6.	$(P \& P) \vdash P$	Thm 7.1
7.	$\oslash \oslash P \vdash P$	5, 6 Cut

7.43

1.	$(((P \rhd P) \& \oslash P) \rhd \oslash P) \vdash ((P \rhd P) \rhd (P \triangledown \oslash P))$	Prem
2.	$\oslash P \vdash \oslash P$	Taut
3.	$(P \rhd P), \oslash P \vdash \oslash P$	2 Thin
4.	$((P \rhd P) \& \oslash P) \vdash \oslash P$	3 And
5.	$\vdash (((P \rhd P) \& \oslash P) \rhd \oslash P)$	4 Ded
6.	$\vdash ((P \rhd P) \rhd (P \triangledown \oslash P))$	1, 5 Cut
7.	$((P \rhd P) \rhd (P \triangledown \oslash P)), (P \rhd P) \vdash (P \triangledown \oslash P)$	MP
8.	$(P \rhd P) \vdash (P \triangledown \oslash P)$	6, 7 Cut
9.	$\vdash (P \rhd P)$	Ex 7.6
10.	$\vdash (P \triangledown \oslash P)$	8, 9 Cut

7.45

1. $(P\&Q) \vdash P$ Thm 7.1
2. $(P\&Q) \vdash Q$ Thm 7.2
3. $\oslash P \vdash \oslash(P\&Q)$ 1 Ex 7.10
4. $\oslash Q \vdash \oslash(P\&Q)$ 2 Ex 7.10
5. $(\oslash P \triangledown \oslash Q) \vdash \oslash(P\&Q)$ 3, 4 Or

7.47 The point is that the negation we defined in Chap. 1 has classical properties that will be inherited by '$\oslash$'. To work this out in detail, we might reproduce the argument we gave in support of Proposition 7.5. We begin with the odd assumption that $f(\mathfrak{a}) = |$ and $f(\mathfrak{a}) \neq |$. We just to want to see what the monks will infer from this. By Exercise 1.16 and Definition 1.8, $id(f(\mathfrak{a}), |) = ||$ and $id(f(\mathfrak{a}), |) = |$ and, hence, $| = ||$. From this, the monks infer $\perp$. So the following is a monastically correct sequent:

$$f(\mathfrak{a}) = |, \ f(\mathfrak{a}) \neq | \ \vdash \perp .$$

But, by Exercise 7.17, this yields

$$f(\mathfrak{a}) \neq | \ \vdash \ \oslash(f(\mathfrak{a}) = |)$$

and, hence, as in the proof of Proposition 7.5

$$\oslash \oslash (f(\mathfrak{a}) = |), f(\mathfrak{a}) \neq | \ \vdash \perp .$$

So, by Theorem 7.5,

$$\oslash \oslash (f(\mathfrak{a}) = |) \ \vdash \ (f(\mathfrak{a}) \neq | \ \triangleright \perp)$$

and, hence,

$$\oslash \oslash (f(\mathfrak{a}) = |) \ \vdash \ -(-(f(\mathfrak{a}) = |)).$$

Now suppose $-(-(f(\mathfrak{a}) = |))$. That is, $id(id(f(\mathfrak{a}), |), |) = |$. By Exercise 1.18, this implies that

$$id(f(\mathfrak{a}), |) + | = |||$$

and, hence, $id(f(\mathfrak{a}), |) = ||$. From this, the monks infer that $f(\mathfrak{a}) = |$. So the following is a monastically correct sequent

$$-(-(f(\mathfrak{a}) = |)) \ \vdash \ f(\mathfrak{a}) = |$$

and, hence, so is

$$\oslash \oslash (f(\mathfrak{a}) = |) \ \vdash \ f(\mathfrak{a}) = | .$$

7.49

1.	$(P \supset Q) \vdash (P \rhd Q)$	Prop 7.9
2.	$((P \rhd Q) \rhd P), (P \rhd Q) \vdash Q$	MP
3.	$((P \rhd Q) \rhd P), (P \supset Q) \vdash Q$	1, 2 Cut
4.	$((P \rhd Q) \rhd P) \vdash ((P \supset Q) \supset Q)$	3 Prop 7.7

7.51

1.	$P, \rightarrow P \vdash \bot$	Prop 7.12
2.	$\rightarrow P \vdash \oslash P$	1 Ex 7.17
3.	$\oslash \oslash P \vdash \oslash \rightarrow P$	2 Ex 7.10

7.53

1.	$\oslash \oslash P \vdash \oslash \rightarrow P$	Ex 7.51
2.	$\oslash \oslash P \vdash \oslash \rightarrow \rightarrow P$	Ex 7.52
3.	$\oslash \oslash P, \rightarrow P \vdash \bot$	1 Ex 7.17
4.	$\oslash \oslash P, \rightarrow \rightarrow P \vdash \bot$	2 Ex 7.17
5.	$\bot \vdash P$	EFQ
6.	$\oslash \oslash P, \rightarrow P \vdash P$	3, 5 Cut
7.	$\oslash \oslash P, \rightarrow \rightarrow P \vdash P$	4, 5 Cut
8.	$P \vdash P$	Taut
9.	$\oslash \oslash P, P \vdash P$	8 Thin
10.	$\oslash \oslash P \vdash P$	6, 7, 9 Prop 7.13

7.55

1.	$(P \rhd \oslash Q) \vdash_I (Q \rhd \oslash P)$	Ex 7.11
2.	$(P \rhd \oslash Q), Q \vdash_I \oslash P$	1 Ex 7.7
3.	$\oslash P, \oslash \oslash P \vdash_I \bot$	Ex 7.20
4.	$(P \rhd \oslash Q), \oslash \oslash P, Q \vdash_I \bot$	2, 3 Cut
5.	$(P \rhd \oslash Q), \oslash \oslash P \vdash_I \oslash Q$	4 Ex 7.17

7.57

1.	$\Phi \cup \{(\phi \triangledown \oslash \phi)\} \vdash_I \oslash \psi$	Prem
2.	$\Phi \vdash_I ((\phi \triangledown \oslash \phi) \rhd \oslash \psi)$	1 Thm 7.6
3.	$\{((\phi \triangledown \oslash \phi) \rhd \oslash \psi)\} \vdash_I \oslash \psi$	Ex 7.56
4.	$\Phi \vdash_I \oslash \psi$	2, 3 Cut

7.59

1.	$\vdash_I \forall x\forall y(Rxy \triangledown \oslash Rxy)$	Ax 7.14
2.	$\forall x\forall y(Rxy \triangledown \oslash Rxy) \vdash_I \forall y(Rxy \triangledown \oslash Rxy)$	Ax 7.12
3.	$\vdash_I \forall y(Rxy \triangledown \oslash Rxy)$	1, 2 Cut
4.	$\forall y(Rxy \triangledown \oslash Rxy) \vdash_I (Rxy \triangledown \oslash Rxy)$	Ax 7.12
5.	$\vdash_I (Rxy \triangledown \oslash Rxy)$	3, 4 Cut
6.	$Rxy \vdash_I (\oslash \oslash Rxy \triangleright Rxy)$	Int
7.	$\oslash Rxy, \oslash \oslash Rxy \vdash_I Rxy$	Ex 7.20
8.	$\oslash Rxy \vdash_I (\oslash \oslash Rxy \triangleright Rxy)$	7 Thm 7.5
9.	$(Rxy \triangledown \oslash Rxy) \vdash_I (\oslash \oslash Rxy \triangleright Rxy)$	6,8 Or
10.	$\vdash_I (\oslash \oslash Rxy \triangleright Rxy)$	5, 9 Cut
11.	$\vdash_I \forall y(\oslash \oslash Rxy \triangleright Rxy)$	10 Ax 7.13
12.	$\vdash_I \forall x\forall y(\oslash \oslash Rxy \triangleright Rxy)$	11 Ax 7.13

7.61

1.	$\{\phi\} \vdash_I \psi$	Prem
2.	$\{\forall\alpha\,\phi\} \vdash_I \phi$	Ax 7.12
3.	$\{\forall\alpha\,\phi\} \vdash_I \psi$	1, 2 Cut
4.	$\{\forall\alpha\,\phi\} \vdash_I \forall\alpha\,\psi$	3 Ax 7.13

7.63 If ϕ has length 0, then it is of the form $R\alpha\beta$ and we have to show:

$$\{\oslash \oslash R\alpha\beta\} \vdash_I R\alpha\beta.$$

This follows from Exercise 7.59. If ϕ is a negation $\oslash\psi$, we have to show:

$$\{\oslash \oslash \oslash i(\psi)\} \vdash_I \oslash i(\psi).$$

This follows from Exercise 7.15. If ϕ is a conjunction $(\psi_1 \& \psi_2)$, we have to show:

$$\{\oslash \oslash (i(\psi_1) \& i(\psi_2))\} \vdash_I (i(\psi_1) \& i(\psi_2)).$$

We reason as follows. (The premises on lines 4 and 5 are our inductive hypotheses.)

1.	$\{\oslash \oslash (i(\psi_1) \& i(\psi_2))\} \vdash_I (\oslash \oslash i(\psi_1) \& \oslash \oslash i(\psi_2))$	Ex 7.60
2.	$\{(\oslash \oslash i(\psi_1) \& \oslash \oslash i(\psi_2))\} \vdash_I \oslash \oslash i(\psi_1)$	Thm 7.1
3.	$\{(\oslash \oslash i(\psi_1) \& \oslash \oslash i(\psi_2))\} \vdash_I \oslash \oslash i(\psi_2)$	Thm 7.2
4.	$\{\oslash \oslash i(\psi_1)\} \vdash_I i(\psi_1)$	Prem
5.	$\{\oslash \oslash i(\psi_2)\} \vdash_I i(\psi_2)$	Prem
6.	$\{(\oslash \oslash i(\psi_1) \& \oslash \oslash i(\psi_2))\} \vdash_I i(\psi_1)$	2, 4 Cut
7.	$\{(\oslash \oslash i(\psi_1) \& \oslash \oslash i(\psi_2))\} \vdash_I i(\psi_2)$	3, 5 Cut
8.	$\{(\oslash \oslash i(\psi_1) \& \oslash \oslash i(\psi_2))\} \vdash_I (i(\psi_1) \& i(\psi_2))$	6, 7 Ex 7.3
9.	$\{\oslash \oslash (i(\psi_1) \& i(\psi_2))\} \vdash_I (i(\psi_1) \& i(\psi_2))$	1, 8 Cut

If ϕ is a conditional $(\psi_1 \rhd \psi_2)$, we have to show:

$$\{\oslash \oslash \oslash (i(\psi_1) \& \oslash i(\psi_2))\} \vdash_I \oslash(i(\psi_1) \& \oslash i(\psi_2)).$$

This follows from Exercise 7.15. If ϕ is a disjunction $(\psi_1 \bigtriangledown \psi_2)$, we have to show:

$$\{\oslash \oslash \oslash (\oslash i(\psi_1) \& \oslash i(\psi_2))\} \vdash_I \oslash(\oslash i(\psi_1) \& \oslash i(\psi_2)).$$

This follows from Exercise 7.15. If ϕ is a universal generalization $\forall \alpha\, \psi$, we have to show:

$$\{\oslash \oslash \forall \alpha\, i(\psi)\} \vdash_I \forall \alpha\, i(\psi).$$

We reason as follows. (The premise on line 3 is our inductive hypothesis.)

1. $\{\oslash \oslash \forall \alpha\, i(\psi)\} \vdash_I \forall \alpha \oslash \oslash i(\psi)$		Ex 7.62
2. $\{\forall \alpha \oslash \oslash i(\psi)\} \vdash_I \oslash \oslash i(\psi)$		Ax 7.12
3. $\{\oslash \oslash i(\psi)\} \vdash_I i(\psi)$		Prem
4. $\{\oslash \oslash \forall \alpha\, i(\psi)\} \vdash_I \oslash \oslash i(\psi)$		1, 2 Cut
5. $\{\oslash \oslash \forall \alpha\, i(\psi)\} \vdash_I i(\psi)$		3, 4 Cut
6. $\{\oslash \oslash \forall \alpha\, i(\psi)\} \vdash_I \forall \alpha\, i(\psi)$		5 Ax 7.13

7.65

1.	$\{\oslash(i(\phi) \& \oslash i(\psi))\} \vdash_I \oslash \oslash (i(\phi) \rhd i(\psi))$	Ex 7.24
2.	$\{\oslash \oslash (i(\phi) \rhd i(\psi))\} \vdash_I (i(\phi) \rhd \oslash \oslash i(\psi))$	Ex 7.19
3.	$\{\oslash(i(\phi) \& \oslash i(\psi))\} \vdash_I (i(\phi) \rhd \oslash \oslash i(\psi))$	1, 2 Cut
4.	$\{\oslash(i(\phi) \& \oslash i(\psi)), i(\phi)\} \vdash_I \oslash \oslash i(\psi)$	3 Ex 7.7
5.	$\{\oslash \oslash i(\psi)\} \vdash_I i(\psi)$	Ex 7.63
6.	$\{\oslash(i(\phi) \& \oslash i(\psi)), i(\phi)\} \vdash_I i(\psi)$	4, 5 Cut
7.	$\{\oslash(i(\phi) \& \oslash i(\psi))\} \vdash_I (i(\phi) \rhd i(\psi))$	6 Thm 7.5

7.67

1.	$\Phi \cup \{i(\phi)\} \vdash i(\chi)$	Prem
2.	$\Phi \cup \{i(\psi)\} \vdash i(\chi)$	Prem
3.	$\Phi \cup \{\oslash i(\chi)\} \vdash \oslash i(\phi)$	1 Ex 7.10
4.	$\Phi \cup \{\oslash i(\chi)\} \vdash \oslash i(\psi)$	2 Ex 7.10
5.	$\Phi \cup \{\oslash i(\chi)\} \vdash (\oslash i(\phi) \& \oslash i(\psi))$	3, 4 Ex 7.3
6. $\Phi \cup \{\oslash(\oslash i(\phi) \& \oslash i(\psi))\} \vdash \oslash \oslash i(\chi)$		5 Ex 7.10
7.	$\{\oslash \oslash i(\chi)\} \vdash_I i(\chi)$	Ex 7.63
8. $\Phi \cup \{\oslash(\oslash i(\phi) \& \oslash i(\psi))\} \vdash i(\chi)$		6, 7 Cut

7.69 Suppose we have a classical derivation of $\varXi \vdash_C \psi$. If the derivation is of length 1, we just apply Exercise 7.68. If the derivation is longer, Exercise 7.68 lets us focus on Axioms 7.2, 7.3, 7.4, 7.5, 7.11, and 7.13. Suppose the last line of the classical derivation features an application of Axiom 7.2 (Thin). Then $\mathcal{E}$ is of the form $\Phi \cup \{\phi\}$ and the sequent $\Phi \vdash_C \psi$ appears on an earlier line. Let $i[\Phi]$ be the result of applying the interpretation function i to each member of Φ. By our inductive hypothesis, there is an intuitionist derivation of $i[\Phi] \vdash_I i(\psi)$. An application of Thin yields $i[\Phi] \cup \{i(\phi)\} \vdash_I i(\psi)$, as desired. Now suppose the last line of the classical derivation features an application of Axiom 7.3 (Cut). Then $\mathcal{E}$ is of the form $\Phi \cup \Psi$ and sequents $\Psi \vdash_C \chi$ and $\Phi \cup \{\chi\} \vdash_C \psi$ appear on earlier lines. By our inductive hypothesis, there are intuitionist derivations of $i[\Psi] \vdash_I i(\chi)$ and $i[\Phi] \cup \{i(\chi)\} \vdash_I i(\psi)$. So, by Cut, there is an intuitionist derivation of $i[\Phi \cup \Psi] \vdash_I i(\psi)$, as desired. Suppose the last line of the classical derivation features an application of the left–right direction of Axiom 7.4 (And). Then $\mathcal{E}$ is of the form $\{(\phi_1 \& \phi_2)\}$ and the sequent $\{\phi_1, \phi_2\} \vdash_C \psi$ appears on an earlier line. By our inductive hypothesis, there is an intuitionist derivation of $\{i(\phi_1), i(\phi_2)\} \vdash_I i(\psi)$. So, by And, there is an intuitionist derivation of $\{(i(\phi_1) \& i(\phi_2))\} \vdash_I i(\psi)$. By the definition of the function i, $\{i(\phi_1 \& \phi_2)\} \vdash_I i(\psi)$, as desired. A similar argument applies to the right–left direction of Axiom 7.4. Suppose the last line of the classical derivation features an application of Axiom 7.5 (Ded) yielding the sequent $\emptyset \vdash_C (\phi \triangleright \psi)$. Then the sequent $\{\phi\} \vdash_C \psi$ appears on an earlier line. By our inductive hypothesis, there is an intuitionist derivation of $\{i(\phi)\} \vdash_I i(\psi)$. So, by Ded, there is an intuitionist derivation of $\emptyset \vdash_I (i(\phi) \triangleright i(\psi))$. Exercise 7.64 lets us derive $\emptyset \vdash_I i(\phi \triangleright \psi)$, as desired. Suppose the last line of the classical derivation features an application of the left–right direction of Axiom 7.11 (Or) yielding the sequent $\Phi \cup \{\phi\} \vdash_C \chi$. Then the sequent $\Phi \cup \{\phi \triangledown \psi\} \vdash_C \chi$ appears on an earlier line. By our inductive hypothesis, there is an intuitionist derivation of $i[\Phi] \cup \{i(\phi \triangledown \psi)\} \vdash_I i(\chi)$. Exercise 7.66 lets us derive $i[\Phi] \cup \{i(\phi)\} \vdash_I i(\chi)$, as desired. Suppose the last line of the classical derivation features an application of the right–left direction of Axiom 7.11 (Or) yielding the sequent $\Phi \cup \{\phi \triangledown \psi\} \vdash_C \chi$. Then the sequents $\Phi \cup \{\phi\} \vdash_C \chi$ and $\Phi \cup \{\psi\} \vdash_C \chi$ appear on earlier lines. By our inductive hypothesis, there are intuitionist derivations of $i[\Phi] \cup \{i(\phi)\} \vdash_I i(\chi)$ and $i[\Phi] \cup \{i(\psi)\} \vdash_I i(\chi)$. So, by Exercise 7.67, there is an intuitionist derivation of $i[\Phi] \cup \{i(\phi \triangledown \psi)\} \vdash_I i(\chi)$, as desired. Suppose the last line of the classical derivation features an application of Axiom 7.13 yielding the sequent $\Phi \vdash_C \forall \alpha \, \phi$ with α not occurring free in any member of Φ. Then the sequent $\Phi \vdash_C \phi$ appears on an earlier line. By our inductive hypothesis, there is an intuitionist derivation of $i[\Phi] \vdash_I i(\phi)$. Inspection of the definition of the function i shows that the variable α does not occur free in any member of $i[\Phi]$. So, by Axiom 7.13, there is an intuitionist derivation of $i[\Phi] \vdash_I \forall \alpha \, i(\phi)$. The definition of i assures us that $i[\Phi] \vdash_I i(\forall \alpha \, \phi)$, as desired.

References

1. Curry, H. B. (1942). The inconsistency of certain formal logics. *Journal of Symbolic Logic, 7,* 115–117.
2. Glivenko, V. (1929). Sur quelques points de la logique de M. Brouwer. *Académie royale de Belgique, Bulletin de la classe des sciences, 15,* 183–188.

3. Gödel, K. (1933). Zur intuitionistischen Arithmetik und Zahlentheorie. *Ergebnisse eines mathematischen Kolloquiums, 4,* 34–38.
4. Gödel, K. (1986). *Collected Works* (Vol. 1). New York: Oxford University Press.
5. Mancosu, P. (Ed.). (1998). *From Brouwer to Hilbert.* New York: Oxford University Press.
6. Martin, N. M. (1989). *Systems of logic.* Cambridge: Cambridge University Press.
7. Peregrin, J. (2008). What is the logic of inference? *Studia Logica, 88,* 263–294.
8. Pollard, S. (2002). The expressive truth conditions of two-valued logic. *Notre Dame Journal of Formal Logic, 43,* 221–230.
9. Pollard, S. (2005). The expressive unary truth functions of n-valued logic. *Notre Dame Journal of Formal Logic, 46,* 93–105.
10. Pollard, S. (2006). Expressive three-valued truth functions. *Australasian Journal of Logic, 4,* 226–243.

Index

Symbols
$(F \cap G)$, 147
$(F \cup G)$, 147
$(F \setminus G)$, 147
0, 127, 147
$<$, 131
Py, 64, 108
$S(F, G)$, 147
$S(\mathfrak{a})$, 7
Sy, 64, 110, 136
$V(\alpha)$, 108
$V(\omega)$, 85
$V(\omega + \omega)$, 86, 87
$V(\omega + n)$, 88
$V(n)$, 76
$\#F$, 123, 145
Π_1^0, 24
$\alpha(x, y, z)$, 152
$\alpha + 1$, 108
$\approx$, 125
$\triangledown$, 175
$\blacksquare$, 172
$\emptyset$, 59, 103, 147
$\equiv$, 143
$\in$, 102, 144
$\leq$, 131
$\mathfrak{F}$, 148
$\mathfrak{F}_\prec$, 149
$\mathfrak{N}$, 126, 127
$\mathfrak{U}$, 127
ω, 85, 113, 148
$\omega + \omega$, 86
$\omega + n$, 88
$\omega_{<y}$, 152
$\omega_{\times y}$, 152
$\oslash$, 170
$\rightarrow$, 167

$\perp$, 170
$\pi(x, y, z)$, 152
$\prec$, 148
$\precapprox$, 150
$\rho(x)$, 77, 90, 102
σwz, 128
σ-closed, 130
$\sigma(xy)$, 5
$\subset$, 148
$\subseteq$, 105, 148
$\tau(x)$, 3
$\upsilon(xy)$, 4
$\vDash$, 40
$\triangleright$, 166
$\vdash$, 39, 162
$\vdash_C$, 181, 185
$\vdash_I$, 181, 185
$\bigcup x$, 73, 108
$\&$, 164

A
Abstraction, 143
Ackermann, W., 64, 66, 67, 73, 85, 86
Agreement, 91
Axiom, 37

B
Basic Law V (BLV), 141–143
Bennett, D., 63
Bernays, P., 28
Biconditional, 107, 124
Boolos, G. S., 123
Burali-Forti, C., 115
Burgess, J. P., 26, 135

S. Pollard, *A Mathematical Prelude to the Philosophy of Mathematics*,
DOI: 10.1007/978-3-319-05816-0,
© Springer International Publishing Switzerland 2014

C

Cantor, G., 145
Cardinal equivalence, 143, 146
Classical equivalence, 183
Compactness, 41
Completeness, 40, 93
Comprehension, 126, 127
Conditional, 166
Conjunction, 19, 124, 164
Connectedness, 69, 71, 92, 105
Contraposition, 171
Corner quotes, 36
Curry, H. B., 20, 188

D

Decidable relation, 185
Dedekind, R., 35
Deduction theorem, 166
Deiser, O., 57
Denseness, 93
Discrimination, 91
Disjunction, 17, 124, 175
Double negation, 17

E

Edwards, H. M., 57
EFQ, 170
Entry, 59
Epsilon-induction, 103
Equiconsistency, 111
Equinumerosity, 125
Equivalence class, 145
Equivalence relation, 26, 143, 145
Extensional equivalence, 143

F

Feferman, S., 50
Formula, 36, 124, 141, 185
Fraenkel, A., 118
Frege Arithmetic (FA), 123, 124, 126, 127
Frege, G., 39–41, 61, 118, 123, 126, 141, 143
Frege-precursor, 61, 109
Function
    characteristic, 15, 44
    interpretation, 183
    one-place, 44
    primitive recursive, 20, 44

G

Gödel number, 47

G

Gödel, K., 29, 40, 45, 47, 183, 187
Gentzen, G., 50
George, A., 47, 123
Glivenko, V., 181, 183
Grelling, K., 18

H

Hallett, M., 69, 118
Heck, R. G., 123
Hereditary property, 11, 38
Heyting, A., 174, 186
Higher rank, 88, 98
Hilbert, D., 1, 24, 25, 50
Hume's principle (HP), 126, 141, 143–145
Hume, D., 126

I

Identity criterion, 3
Incompleteness
    expressive, 43
    proof-theoretic, 48
Inconsistency, 43
Independence, 107
Induction, 11, 38, 68, 70, 89, 98, 103, 112,
    132, 148
Inductive hypothesis, 12, 39, 104
Interpretability, 101, 111
Interpretation, 39, 106, 109
Intuitionist equivalence, 183
Irreflexivity, 69, 92, 104

K

Kanamori, A., 59, 118
Kaye, R., 64
Kunen, K., 118

L

Landini, G., 59
Law of excluded middle (LEM), 18, 176, 185
List(s)
    P-closed, 70
    S-closed, 67
    -token, 56, 75
    -type, 56, 57
    as numbers(s), 67
    blank, 59
    cofinite, 91
    entry-closed, 60
    finite, 72
    formation of, 59, 60

hereditarily finite, 72–74
of bounded rank, 78, 90
pure, 63
unranked, 56, 57
well-founded, 71
Zermelian, 90
Logic
first-order, 39, 184
intuitionist, 167
second-order, 145
Logical consequence, 40

M

Mancosu, P., 24, 25, 28, 181
Martin, N. M., 171, 184
Mirimanoff, D., 69, 105
Model, 39, 106
Monadic Frege Arithmetic (MFA), 145

N

Nagel, E., 47
Newman, J. R., 47
Number
as list, 66
natural, 37, 132
real, 95
simpliciter, 132
standard, 41
super-, 70–72
super-duper-, 88
Numeral
-token, 2
-type, 2

O

Order type, 118
Ordinal, 118
Outranking, 102

P

Pairing, 125
Part, 64
Peano arithmetic (PA), 35, 135
Peano, G., 35
Peirce's law, 174, 177
Peirce, C. S., 174
Peregrin, J., 161
Pollard, S., 55, 102, 115, 178
Precedence, 102
Precursor, 60, 109

Predecessor, 10, 113
Primitive Recursive Arithmetic (PRA), 1
Proof, 39

Q

Quantifier
existential, 19, 124
numerical, 124
universal, 20
Quine, W. V. O., 36, 143

R

Rang, B., 59, 143
Rank, 75, 77, 89, 104
Recursive definition, 10
Reflexivity, 131, 143
Relative consistency, 111
Representability, 44

S

Sentence, 37, 124, 181
Sequent, 162
Set(s)
closed under set formation, 103
finite, 148
formation of, 102
hereditarily finite, 114
member-closed, 109
of bounded rank, 102
transitive, 105
Shapiro, S., 144
Skolem, T., 118
Soundness, 41
Standard
model, 46
number, 41
Subset, 105
Successor, 6, 7, 37, 70, 128
Summation, 18
Symmetry, 143

T

Testability, 177
Theorem, 37, 172, 183
Thomas, W., 59, 143
Transitivity, 69, 92, 104, 131, 143
Truth table, 17

U

Undecidability, 46

Unsatisfiability, 43
Upper bound, 93
Uzquiano, G., 90

V
Van Aken, J., 102
Van Heijenoort, J., 39, 45, 61, 74, 115, 118,
    143
Variable
    bound occurrence, 36, 124, 185
    free occurrence, 36, 124, 185
    object, 123
    property, 123
    relation, 123

Velleman, D. J., 47, 123
Visser, A., 152

W
Wang, H., 35
Weir, A., 144
Well-foundedness, 69, 71, 104
Well-ordering, 70
Weyl, H., 115
Wong, T. L., 64

Z
Zermelo, E., 59, 74, 86, 90, 108

Printed in the United States
By Bookmasters